Medical Technics

Forerunners: Ideas First

Short books of thought-in-process scholarship, where intense analysis, questioning, and speculation take the lead

FROM THE UNIVERSITY OF MINNESOTA PRESS

Don Ihde
Medical Technics

Jonathan Beecher Field
Town Hall Meetings and the Death of Deliberation

Jennifer Gabrys
How to Do Things with Sensors

Naa Oyo A. Kwate
Burgers in Blackface: Anti-Black Restaurants Then and Now

Arne De Boever
Against Aesthetic Exceptionalism

Steve Mentz
Break Up the Anthropocene

John Protevi
Edges of the State

Matthew J. Wolf-Meyer
Theory for the World to Come: Speculative Fiction and Apocalyptic Anthropology

Nicholas Tampio
Learning versus the Common Core

Kathryn Yusoff
A Billion Black Anthropocenes or None

Kenneth J. Saltman
The Swindle of Innovative Educational Finance

Ginger Nolan
The Neocolonialism of the Global Village

Joanna Zylinska
The End of Man: A Feminist Counterapocalypse

Robert Rosenberger
Callous Objects: Designs against the Homeless

(Continued on page 84)

Medical Technics

Don Ihde

University of Minnesota Press
MINNEAPOLIS
LONDON

"Sonifying Science: Listening to Cancer" previously appeared in *Nursing Philosophy* 18 (2017), wileyonlinelibrary.com. "Aging: I Don't Want to Be a Cyborg, I and II" and "Aging Cyborg, III, IV, V, VI, and VII" previously appeared in a different form in "Aging: I Don't Want to Be a Cyborg," *Ironic Technic* (Automatic Press, 2008), reprinted here with permission of Automatic Press and Vincent Hendricks. "From Embodiment Skills in Computer Games to Nintendo Surgery" previously appeared as "Embodiment," in *Debugging Game History: A Critical Lexicon,* ed. Henry Lowood and Raiford Guins (MIT Press, 2016); reprinted with the permission of MIT Press. "We Make Technology, Technology Makes Us" previously appeared on GuernicaMag.com December 1, 2017; published by permission of Daisy Alioto.

Published by the University of Minnesota Press, 2019
111 Third Avenue South, Suite 290
Minneapolis, MN 55401–2520
http://www.upress.umn.edu

The University of Minnesota is an equal-opportunity educator and employer.

Contents

Introduction

THIS IS A BOOK about aging and medical technologies in the twenty-first century. It will follow a style that evolved in this now sixth of my books with "technics" in the title. On one side personal, it is also autobiographical in that it will be my own experience of aging that is being described; on the other side I am a philosopher of technology, a recently arrived type of philosophy that emphasizes the role of technologies in many dimensions of human life, both individual and social. And, finally, my approach is known today as *postphenomenology,* a modified type of phenomenological analysis specifically aimed at what is unique about technics, or the materiality of instruments and tools developed in medical science.

The origins of this inquiry are distinct to the twenty-first century, specifically 2008, which was a major turn in my own life, when aging became dramatic. I am now eighty-five years old; 2008 was my seventy-fourth year, and this book spans ten years of aging and the medical procedures that I underwent during that time. In some ways I was a typical aging male. In my role as a philosopher of technology, I had just completed three trips to China, a country that is intensely involved in what I will call a postmodern technological revolution. Today its ambitions include becoming the world leader in AI and many other fields. I had just made three lecture tours to China in 2004, 2006, and 2007, and while there, visiting ancient sites such as gardens up a steep mountain, the Great Wall, and ancient cities, I found myself panting, tiring, and having difficulty maneuvering for the first time in an active life. After my first bad short-breathed experience back in the United States in 2006, I went to see a cardiologist at my university hospital. His diagnosis, revealed in an angiogram, was cardiovascular disease. I underwent balloon angioplasty plus a stent in my most blocked artery. (I summarize all the medical procedures I experienced in "Aging: I Don't Want to Be a Cyborg, I and II" and "Aging Cyborg, III, IV, V, VI, and VII.")

Now for "Technics": I borrowed this term from Lewis Mumford, whose *Technics and Civilization* (1934) was the best-known intellectual history of technology when I introduced a new course in philosophy at Stony Brook University shortly after my arrival in 1969. He was not at the time thought of as a philosopher, but the term "technics" seemed to nicely cover both the sense of technologies in use and the materiality of technologies, which I found helpful. The early seventies marked my first turn to philosophy of technology, and in background reading I became aware that many European technology theorists were using a vocabulary that included "technology" and "technique," including early writers such as Foucault, Ellul, and Marcuse. In each case they seemed to turn everything into a technology that some regarded as a carryover from the materiality of technologies to analytic techniques such as language. I was skeptical of this expansion, although I could see how imaged techniques such as sports motility and even sexual praxis might indirectly reflect as "techniques" inspired by "technologies" without actually being technologies. I also became aware of how the tone of many early philosophers of technology, such as Heidegger, Ortega y Gasset, and Jaspers, often cast a dystopian shadow upon technologies viewed overall. And later Mumford also became critical in his *Myth of the Machine* (1970). His earlier *Technics and Civilization*, however, cast a more objective view upon the important role of technics in the history of civilization and recognized that the social impact of the clock paved the way for the Industrial Revolution. For me, technologies remain material while "hardware," concrete, and technics resist totalization, a sense that I wish to continue in *Medical Technics*. I grew up on a Kansas farm, and its culture included lots of "junk" technologies. Every farmer had a discard pile of technology parts from worn-out, abandoned machines. These were often later used for invented "bricolage" machines that my brother, father, and I would invent once all of us developed the skill of welding. Much later, while first doing an MDiv and then a PhD in the Boston area from 1956 to 1964, I had become a chaplain at MIT (1958–1964) and spent most of my lunch hours in the faculty and graduate dining room just two doors away from my office. There I met many of MIT's engineer-inventors only to find that many had midwestern heritages with the bricolage technologies I had known as a boy. This was in the late fifties, early sixties when "planned obsolescence" was popular in commerce, and many engineer-inventors admitted that their sponsoring companies often encouraged them to produce compound parts that, despite having shorter shelf lives, were much cheaper. I later wondered about the countermove with joint implants for which the problem was precisely shelf life. Early knee replacements lasted only about six to seven years, and thus an athletic person in his or her mid-thirties could have to undergo surgery again several times in later life. In this case the implant technologies had improved enough that when my own replacement knee implant was surgically placed in 2013, I, then seventy-nine, could expect at least fifteen or more years of shelf life. From all this it seemed to me that the technofantasies of technological eternal life simply missed the reality of technological finitude. (If Heidegger's convincing "being toward death" underlined human contingency and finitude, should not technologies also be finite and contingent? Technologies, too, have a being toward death as shelf life.)

If my early experiences of medical technics again begin with my seventies, I sometimes think, *what if I had been unlucky enough to have been born a century earlier?* I think the answer would have to be I'd likely be *dead*. We now take careful note of how longevity has changed dramatically, at least in all industrialized countries. Human lifespans a century ago were forty-five or a little more. Today at least ten countries have about eighty-five as an average lifespan. How did this happen? First, a major factor is sanitation. Until national practices turned to sanitation, diseases like typhoid, the earlier plagues of the world, and other epidemics could kill vast proportions of societies. But, now obvious in retrospect, until we could recognize the microorganisms—germs, bacteria, and viruses—that caused diseases, there could be no informed therapies. Let me, somewhat oversimply, point out that the optical imaging of early modern science played a major role. Leeuwenhoek and many others, including Galileo, developed the microscope, which in turn revolutionized our view of a microworld, including reproduction, with the visual discovery of sperm and eggs and, later, disease-causing microorganisms. My own 2008 heart repair (not replacement) procedure could not have been diagnosed or surgically undergone a century ago. The wall of my surgeon's imaging displays at my pre-surgical consultation, with his moving cursor to show what was broken, could not have existed until the twenty-first century.

So, if we return to our century, we see a postmodern world where we are undergoing what I sometimes call a "second scientific revolution," very apparent in its medical dimension and led by a turn to laparo-, micro-, and nanoprocesses, especially in imaging technologies and surgical procedures. Let us look at some of the major changes even from the twentieth century. Surgery, for example, has clearly moved from very large processes—akin to butchering—to the laparoscopic processes of today. Even appendicitis once necessitated a major operation. Yet when my wife found she had bone spurs in her shoulders, inhibiting tennis playing, laparoscopic surgery left at most a half-inch scar that, when she added a few weeks of physical therapy, allowed play to resume; or, as will be discussed in "Aging Cyborg, III, IV, V, VI, and VII," my cataract surgery was performed with a femtosecond laser process one million times more accurate than previous laser surgery technologies. It took fifteen minutes per eye and yielded what my ophthalmologist terms "perfect vision." This was unheard of in the twentieth century. This procedure is performed on approximately 3.5 million patients per year in the United States.

I opened this entry with the revolution in longevity followed by the switch to microsurgery, which now is frequently performed robotically at a distance. I group these over yet another change from earlier medical technics in "Aging: I Don't Want to Be a Cyborg, I and II" and "Aging Cyborg, III, IV, V, VI, and VII," where I describe my own procedures. These chapter titles point to the growing capacity of medical technics to add permanent technologies to aging bodies. While tooth crowns have a relatively long history, implants constructed of metals and plastics in joints such as the knee, hip, and elbow are relatively new and common, again with close to millions of procedures per year in the United States. Stents and other permanent implants multiply regularly. I decided that it is better being a cyborg than disabled or worse, dead, after all. I chose Donna Haraway's popularized notion of a cyborg as a hybrid mix of human, technology, plus sometimes animal parts over the more mechanized versions of earlier imagined cyborgs influenced by "cybernetics" models thought of by Norbert Wiener (whom I knew at MIT), since there is a clearly different pre- and postcyborg experience. So far, I have mentioned the well-known fact that early philosophy of technology was often dystopian. Yet its opposite exaggeration is often even earlier and hypes a *utopian* technofantasy. This technology will do marvelous things—perhaps even allow eternal life. The cyborg metaphor also includes dreams expressed in notions of the *posthuman,* or worse, *transhuman* fantasies. Posthumans are narrower: they fix upon genuine gains and project enhancement developments. Two common sensory prostheses are visual and auditory. When told by one of my ophthalmologists after cataract surgery that I now had "perfect" vision, I recalled what had been excellent vision that matched African Dogon vision, which was so sharp that the satellites of Jupiter could be seen with naked eye vision (a feat I used to match in my Vermont place). I recognized this result was a restoration to younger sharper vision, not an enhancement. And I wish my state-of-the-art digital hearing aids could do the same, but I continue to reject the possibility of a cochlear implant that does not restore hearing but allows a different hearing altogether, which is very difficult to learn as users testify. Transhumanists prone to many science fiction technofantasies imagine a self-guided human evolution that is *bionic* beyond default. As I point out later, all actual implants attract a biological film that can be "indolent," as mine was fortunate enough to be, or bacteriologically toxic, even deadly. Transhuman, utopian technofantasies portrayed in film and television somehow ignore actual biofilms in transplants. Postphenomenology, respecting biological science, remains skeptical of such technofantasies.

Within the spectrum of *praxis* philosophies now at work on technics, postphenomenology often converges with critical theory, the current version of neo-Marxian philosophy. Many note how the work of Andrew Feenberg and my work converge, but Friedrich Kittler also has played a role. His *Taking Care of*

Nietzsche shows how unexpected outcomes occurred when the typewriter was invented in 1878. Most secretaries were male, proud of their well-honed bodily skills using pens. The shocking invention of the typewriter provoked a "Luddite" reaction in the male secretaries, who thought typing "de-skilled" them and refused to adopt it. Instead, young women "pre-skilled" by piano training responded to an opening to escape the limits of domesticity, and within little more than a decade the vast majority of secretaries were female.

And now, I add a special and philosophical interest. On a philosophy of science side, I have long argued that science would be greatly enriched—in spite of its highly successful use of visualization from imaging technologies—were it to be multisensory, especially regarding auditory and acoustic imaging. I have long been known for this interest, and my *Listening and Voice: A Phenomenology of Sound* (1976), its second edition, *Listening and Voice: Phenomenologies of Sound* (2007), and *Acoustic Technics* (2015) are featured in an acoustic technologies in medicine follow-up, described in "Sonifying Science: Listening to Cancer."

Indeed, in recent science publications there is evidence that sonification is being given as an alternative mode of science reporting. Those interested in the Cassini space probe's last missions can hear dust particles hit the probe as it descends for its last travels that crash onto the Saturnian surface. Or, one can tune in to hear the sonic "moans" of glaciers melting; or follow the new "Superhenge" discovered by subsurface sounding in England, near Stonehenge but five times bigger; or, one of my favorites, learn about the 8,600 buried pyramids in Central Mexico discovered by airplane magnetometer in 1996.

Let's turn back to philosophy of technology, particularly as related to medical technics. As previously noted, my own approach is called postphenomenology. Philosophy of technology, indeed like the timeframe for this book, is a relatively new interest of philosophy. It does have some notable nineteenth-century harbingers, both neo-Hegelians. After all, it was Hegel who got us to think of "philosophies of ______." He spoke of the philosophies of history, religion, and so on, so why not philosophy of technology? Two of his followers, Karl Marx and Ernst Kapp, did take technologies seriously. Kapp, indeed, was the first to title a book *Philosophy of Technology* in 1877. Kapp thought of technologies as analog organs. Far better known was Karl Marx, who inverted Hegel's idealism into a kind of "dialectical materialism" that saw technologies whose change over time shaped the "means of production." Like Mumford, Marx looked at how different technologies produced—in his case deterministically—different societies. But it was only after World War I, with the large military-industrial technologies of tanks, airplanes, and machine guns, that a whole generation of major philosophers began to take technologies seriously in philosophical concern. Of these, Martin Heidegger was—and is—thought of as philosophy of technology's inventor, but Ortega y Gassett, Karl Jaspers, and many others of this first generation also shaped an opening to philosophy of technology, first taken mostly as transcendental or generic technology, often viewed with a dystopian evaluation. A second generation, mostly students of the first, continued the transcendental-dystopian views, now including many neo-Marxists centered in the Frankfurt School: Arendt, Marcuse, Adorno, Horkheimer, and Habermas. By my time, Hans Achterhuis noted in *American Philosophy of Technology: The Empirical Turn* (2001) that there had occurred a turn to the examination of concrete technologies. And while all three generations held, as I term it, a *praxis* or practices emphasis found in a limited number of philosophical traditions (Marxism, phenomenology, pragmatism, process philosophy), now both European and American, the dystopian tone had also changed, and an empirical turn had taken place.

Postphenomenology first took shape largely out of my technoscience research group at Stony Brook and is clearly a technology-focused version of phenomenology. Aided by the "case study" concreteness of STS (science technology studies) and presentation of postphenomenology research panels at all the major STS style conferences, this movement is a major presence in STS venues. My historical mentors were classical phenomenologists: Edmund Husserl, Martin Heidegger, and Maurice Merleau-Ponty, plus Paul Ricoeur with hermeneutics. I was later largely involved in the eighties discussion of "non-foundational" philosophy stimulated by Richard Rorty and leading back to John Dewey, where I found much in pragmatism that helped enhance concern with technologies. I had long felt uncomfortable with the early modern subjectivist tone that Husserl maintained in his dominantly idealist adaptation of Descartes and with both Husserl's and Merleau-Ponty's claims that science was distant from the lifeworld, thus opening the way for the exaggeration that phenomenology was antiscientific. Postphenomenology emphasizes that the already interrelational analysis of human–world relations, which postphenomenology makes into a mutual co-constitutive relation, includes a mediating role for materiality, technics, and technology so that human–world interrelations become human–technics–world relations. The most easily emphasized and grasped version of this modification of phenomenology, I held, could be well illustrated in the case of imaging technologies. If one is familiar with the plethora of similar interrelationary analyses in the mid-twentieth-century arrival of science studies—the new sociologies of science, feminist critiques, and philosophies of technology—one can see that postphenomenology belongs to this intellectual milieu. Briefly put, postphenomenology joins the now widespread notion that human knowledge is perspectival, bodily located, and multisensory—or as I sometimes put it, *praxis-perceptual*. One can see there are many postmodern connections to this particular focus upon technics.

What I found, even from the early days of my turn to philosophy of technology—marked by the publication of my *Technics and Praxis: A Philosophy of Technology* (1979), cited as the first English language book in philosophy of technology—was that what I could do as a philosopher underwent deep changes. For example, with my first (and final career) move to Stony Brook University, I found myself working with the engineering college and collaborating with a wide regional "technological literacy" program supported by the Sloan Foundation. I soon found myself going around the entire state (New York and later New Jersey and Florida), often talking to groups of adults about computer education for very young children. Computer manufacturers were at that time mounting a serious advertising program aimed at stimulating laptop sales for young children. One theme was to claim that unless your child learns computer skills early, the child will "fall behind" and not do well in school. I was already doing heavy research in the history of technologies and their commercial propagation, often in collaboration with Ruth Cowan, a well-known historian of technology at Stony Brook. One of her examples echoed early successful sales campaigns that had to do with washing machine soaps: unless you have the proper washing machine soap, you will get germs and disease—so buy our soap. I found myself urging reasonable caution, telling worried parents that computer prices would come down, children would learn skills if interested anyway, and to not too easily rush into the latest thing. The same applied to worries about screen time: was playing computer games reductive? Was the reduction of gaming to eye–hand coordination bad? In this case, as I learned from so many of my late life doctors, this gaming has turned out to be a kind of pre-skilling for "Nintendo surgery" (see "From Embodiment Skills in Computer Games to Nintendo Surgery") today widespread in medical technics, space exploration, and drone piloting. Pre-skilling turns out to be also related to convergent themes among postphenomenology and critical theory. Much of classical Marxian theory relies on notions of *alienation* in work, often called de-skilling as in my male secretary example. Pre-skilling, unpredicted in the secretary example, softens the notion of alienation. Another variant occurs in the development of early musical synthesizers. Trevor Pinch and Frank Trocco (Trevor a longtime collaborator) published *Analogue Days* (2004) showing that the Moog, by virtue of using a keyboard as its control device, vastly overcame the Buchla, which had a more complex control panel, another example of pre-skilling since many musicians were already familiar with keyboards.

Alongside my U.S. work with the Sloan Foundation and technological literacy, I was also working with Lego in Denmark in an interdisciplinary group worried about how computer games might encourage obesity for lack of bodily exercise. Our group imagined many "Wii-style" games like those that are now prominent, but this exercise also helped me understand more deeply the side issues of pre-skilling for Nintendo surgery and distant sensing noted in "From Embodiment Skills in Computer Games to Nintendo Surgery."

"Postphenomenological Postscript: From Macro- to Microtechnics" returns to philosophical themes and "We Make Technology, Technology Makes Us" concludes with a postretirement interview with a story related to a change in postretirement life. Upon moving into a Manhattan apartment, I began to receive multiple requests for interviews, one of which came from an Israeli techie interested in AI, machine learning, and the like. He invited me to give a presentation on technics in his East Village workshop, and there I met Daisy Alioto, a freelance writer who in turn did the closing interview of how we humans invent technologies, which then also invent us.

Sonifying Science: Listening to Cancer

SCIENCE AND ART have been lifetime passions of mine, and although I have written much more about the former, I have returned to the practice of doing art again in my later life. In the case of science, I have focused upon its material embodiment in technologies, primarily instruments, and drawn from a long *praxis* set of traditions, particularly phenomenology and pragmatism, which I call postphenomenology. I have long argued that without technologies—instruments—there could be no science. And as I have returned to doing art, I realize the same is the case for art practices.

In recent work, I have taken a long, deep historical look at both art and science practices with respect to tool kits and somewhat arbitrarily begun my narratives with the Ice Ages. I now realize that a kind of archeo-art and archeoscience, both practices with tool kits, go at least this far back. Indeed, what is clearly an art tool kit has been found and dated back to 100,000 BP (BP is a science dating convention meaning "before present"), well before the Ice Age, in Blombos Cave in South Africa. It is a paint-mixing kit that uses a giant sea snail shell as the container, ground up charcoal, red and yellow ocher, bone marrow, and other liquid to make a black/red/yellow palette. It is not clear what types of art may have been produced, but small shells with holes and coloring indicate at least necklace making. Similar finds, dated 75,000 BP, also show color applications to various artifacts. Of course, by the early Ice Ages, 45,000 to 15,000 BP, the proliferation of sophisticated cave paintings is well known.[1]

What of science, or what I shall call archeoscience? While nothing that goes back 100,000 years is yet known, by mid–Ice Age, artifacts such as reindeer antlers, bone, and stone items depict marks clearly indicating lunar calendars 36,000 to 22,000 BP. It is well known that our ancestors, pretty much all over the world, were keen observers of the nighttime skies and quantified the movements of various celestial objects to produce lunar and solar calendars, regularized the solstices, named constellations, and used both a recording technology (antlers, bones, stones incised) and standardizing observational devices (gnomons, stone rings, sighting constructions) to regularize this knowledge. In short, Ice Age art and Ice Age science were "technologically embodied."

Of course, as the new philosophies, anthropologies, and sociologies of science recognize, all this practice was culturally embedded in a wider lifeworld of human activity. For hunter-gatherers and very early societies, migration, navigation, and other regularities were needed to stabilize life; for later agricultural societies, crop planting and animal breeding times were needed and knowledge of such natural regularities fit those needs as well. In my own earlier career, I made much about navigational knowledge as well, as applied to both sea voyaging or migratory pathways.

1. Jill Cook, *Ice Age Art: Arrival of the Modern Mind* (London: The British Museum Press, 2013). This catalogue of Ice Age Art was the largest Europe-wide collection of art artifacts to date.

Ancient art and science clearly had to rely on the acuity but also the limits of human perception, which took different shapes in relation to different environments. The astronomy of prehistoric and early history was limited to "eyeball" observation. What marked what we today call "early modern science" was an optical-technological revolution. Galileo used the newly invented *telescope* (but also a *microscope*), which opened up macro- and microrealms of extended perception never previously experienced. Yet optical technologies were actually first more employed in art prior to science.[2] The Renaissance artists used the camera obscura, camera ludica, and various grip frame devices to produce what we today call "Renaissance perspective" and other verisimilitude effects in art productions. Galileo's helioscope for viewing sun spots was a century and a half later.

If "art preceded science" in the use of optic technologies in Renaissance times, then it was paralleled by an equal proliferation of acoustic technologies. All our "classical" instrument types were also invented in this period: violin, cello, sackbut (trombone), cornet, et cetera. And as with the helioscope, Galileo also adapted parts of musical instruments and practices into his science investigations: frets for measurement of inclined plane motions, pulse for timing of the Pisa pendulum observations, and so on. (Note in passing that his father was the inventor of Italian opera and a musician.)

A cultural question arises, however, concerning what becomes a different set of trajectories in art and science. If, as so many of our histories and philosophies of science would have

2. Don Ihde, "Art Precedes Science, or Did the Camera Obscura Invent Modern Science?" in *Instruments in Art and Science,* ed. Helmar Schramann, Ludger Schwarte, and Jan Lazardzig (Berlin: Walter de Gruyter, 2018), 383–93.

it, physics and astronomy were the most refined of early modern sciences, then the use of optical technologies as favored instruments skews those sciences in a visual direction (no early acoustic technologies were relevant to astronomy until radio astronomy in the twentieth century). This may relate to what becomes an increasingly *visualist* preference in science practice. And, of course, the invention of early modern anatomy fits here as well, with da Vinci and Vesalius and their exploded diagram drawings, a visualist style from the Renaissance on. By late modern times, what I have called a sophisticated visual hermeneutics had become virtually the standard for science depiction—see my *Expanding Hermeneutics: Visualism in Science* (1998) for an expanded discussion of this cultural contingency in science.

To this point, from Ice Age art to the Renaissance, one might say that art precedes science in the invention and use of technologies-tools-instruments. By late modernity, which in my estimation begins in the nineteenth century with the arrival of much more complex mega- and microtechnologies, there is a shift in which science begins to develop and use technologies that only later are adapted to art. And with this shift—if I am right—there is a reversal of discovery and use such that often art, later than science, opens up a technology to scientific practice. For example, one of my most recent venues was the sixth Computer Art Congress. Artists did not invent computers, although they did invent many Renaissance optics. I will here focus on what I am broadly calling *the sonification of science* and acoustic technologies. As I pointed out in *Expanding Hermeneutics: Visualism in Science,* since the seventeenth century, science in practice has favored vision as a sort of cultural choice and has developed, particularly in imaging technologies, a very sophisticated visual hermeneutics. However, recently and often due to scientist-musician and other artist hybrid practitioners, sonification has begun to emerge as a major interest in science imaging.

Sonar, Radar, and Early Sonification

In ordinary human experience, we are familiar with echoes, the sound of our voice, and a musical instrument "returning" in a mountain setting. But it was not until the nineteenth-century discovery of what we now call the *electromagnetic spectrum* (EMS) that science found a very wide range of wave phenomena that included what we experience but that also far exceeded what we directly could experience. Radar and sonar are "echo-phenomena" but in their instrumental incarnations utilize wave phenomena that we cannot directly experience—but that, through technological mediation, we can again experience.

Radar and sonar were both early twentieth-century technologies that worked like superecho devices, and their most familiar early use was military, to detect airplanes (radar) and submarines (sonar). Active forms used short, sound-like bursts that would return, echo-like, to a receiver. The receiver could actually be acoustic—one could hear the "pings"—or could be made visual on a video display screen. As common knowledge has it, World War II might well have had a very different outcome had the United States and United Kingdom not had superior radar-sonar technologies. Of course today both systems are vastly improved and used for many kinds of detection, mapping, and imaging (see my *Acoustic Technics,* 2015). As an aside, I note again that, for astronomy, the invention of radio astronomy was the first breakthrough from the optical spectrum, and it is now used to detect radiation not only from dark or nonoptical parts of the sky but to detect the background radiation of the entire universe. Thus, unlike early modern science, today we can both "see" and "hear" the heavens.

Computers and Data-Image Inversions

Above I have hinted that the discovery of the EMS was a major science breakthrough of the nineteenth century. The twentieth-century counterpart, I would contend, is the invention of the computer and its digitalization that drives so many of our practices, both in art and science, today. I shall focus on its capacity to invert data and image. We denizens of the twentieth and twenty-first centuries are all familiar with solar system exploration, one practice that heavily uses the data-image inversion of computer digitality. Cassini is a space probe that has been dedicated to imaging the rings, moons, and other phenomena of our sister planet Saturn. The probe, once in orbit around Saturn, could aim its cameras, radar, or whatever else to the rings. But such dramatic 3-D images are not directly sent back for us to see. Rather, these visual gestalts are first transformed into the familiar binary zeroes and ones of data. These data streams, popularized in such movies as *The Matrix,* are sent by radio processes to a home station where they are again transformed back into the dramatic images we see of Saturn's rings, moons, and so on. This inversion process is used in many, many applications and is a taken-for-granted capacity of what I call postmodern imaging. Today's imaging is thus both compound—images, devices, plus computer tomography—and translational—data into image, image into data.

The Artistic Innovation

What follows is somewhat of a caricature, but it has actually happened. One might say that in a normal science, one could take for granted that data-image inversions are simply translations from data to *visual* images. And of these, we have multitudinous examples. Let us turn now to more medical examples. If we are looking for a brain anomaly, we use MRI, CT scans, PET scans, or fMRI, all visualizations of the tumor or malignancy that then gets tomographically imaged as a 3-D model. Pap smears image on a slide cells, healthy or malignant, in order to detect a cancer in time for treatment. There is a similar process for prostate cancer in males.

I now turn anecdotal. Many know that for several decades I have been especially interested in auditory, acoustic phenomena, and on many lecture trips here is what happens: I introduce sound bites of what I take to be innovative and interesting sounds via a CD and player. When I do this, quite frequently someone in the audience is an actual performance or other artist-practitioner and will present me with an example, say a CD, of his or her own art product. One of my favorite examples is a CD titled *Ground Station.*[3] I shall describe it. When I play the bite, what one hears is a very narrow kind of minimalist digital piano music. The songline consists of a very few notes that sound in what seems to be a random order, reminiscent of a Philip Glass or Steve Reich piece. And it is obvious that the few notes played repeat themselves. What produces this "song," however, is very complex. It is a translation of data that comes from a geostationary satellite such as those that beam the data to run GPS instruments in cars. The voice and visual display in your car, however, do not come from this data stream. Instead, they come from a data stream that corrects satellite wobble. For, like the earth or in fact any object in orbit, motion is not actually smooth but filled with minute wobbles that, if not corrected, would yield a confused pathway for the GPS in your car.

This acoustic result is one of my favorite art examples that can be used to show the culturality of science's visualism. Any musician, even short of a perfect-pitch listener, could quickly be

3. Daniel Joliffe and Jocelyn Robert, *Ground Station,* 2003, Surrey Art Gallery.

taught to recognize what each note means vis-à-vis the position in a wobble of the satellite. The listener could "hear" the angle. Now I shall turn to a similar example, which I call "listening to cancer." I have not been able to discover who was the first person to recognize that sonification could be used to do science. It is clear that a number of performance artists did discover that in the data-image inversion process, one could turn data *either* or *equivalently* into visual or acoustic images. But in the articles that first pointed out the data to acoustic image inversions, it turned out to be artistically or musically trained scientists who made the switch.

Data to Acoustic Imaging

Sonification in science, especially in this era of high-tech imaging instruments, is relatively new and often accidental. But it is occurring with increasing frequency and with often astonishing and innovative results. A recent "Science and Technology" article in the *Economist* (March 19, 2016), "Now Hear This," points to this sporadic history. During World War I, Heinrich Barkhousen used a modified radio-like device to try to listen in to Allied communications—instead he got lots of static that he eventually realized came from lightning storms in the distance.[4] Others have pointed out that this would eventually play a role in radio astronomy as noted above. By mid-century, seismologists realized that if they sped up sound data, they could better discern pre-earthquake patterns. As noted above, space explorations like Cassini, when sonified, created hailstorm-like sounds later identified as debris from space hitting the rings.[5] Later in

4. "Now Hear This," *The Economist*, March 19, 2016, 83.
5. Ron Cowen, "Sound Bytes," *The Scientific American*, March 2015, 45.

the article, one Robert Alexander is cited for his sonification of sun flares to which he listens.[6] His name immediately rang a memory bell for me since an earlier (March 2015) *Scientific American* article "Sound Bytes," had identified him as a pioneer of science sonification.[7] This was the article that first described "listening to cancer."

As it turns out, Robert Alexander is by vocation both a scientist and an artist; he is a composing musician and was previously a graduate student at the University of Michigan.[8] But, like so many new practices in technoscience, sonification has a complex and growing development. Anecdotally, often when I claim that science's visualism is culturally contingent, I get objections from scientists in the audience, and one line of objection claims that we humans are physiologically-neurologically-visually superior. So it is not surprising that the current focus upon THE BRAIN, or its neurology, should be used in this more contemporary context to counter the older visualist claims.

"Sound Bytes" does just that: "Our ears can detect changes in a sound that occur after just a few milliseconds." This is a claim by Andrew King, Oxford University neurologist, who goes on to state that by comparison the eye's limit for detecting a flickering light is about fifty to sixty times a second.[9] Another neurologist, Bechara Saab of the Neuroscience Center Zurich, claims more generically that a mammal's "auditory system is faster at transmitting neural signals than most other parts of the brain. This system holds the largest known connection between neurons, a giant synapse called the calyx of Held. The calyx can release neurotransmitters eight hundred times a second. In contrast, the

6. Cowen, 45.
7. Cowen, 45.
8. Cowen, 46.
9. Cowen, 46.

visual pathway does not have such a speedy neural connection."[10] While I do not claim expertise in this domain, and I definitely do not want to substitute an aural bias for a visual one, the new neurologies are of interest in this scientific-cultural battle.

Now I shall turn to the acoustical diagnosis of cancer—listening to cancer. The current, dominant diagnosis, and I take here cervical cancer as my example, begins with a biopsy from a Pap smear sample made into a slide and visually analyzed by an expert reader of slides, and eventually the result is reported to the patient. One problem, of course, with this type of diagnosis—and with many medical tests—is the delay time that stimulates a period of anxiety for the patient. This delay problem was noted by yet another scientist-artist, Ryan Staples in the United Kingdom, who imagined a much faster process if the diagnosis could be sonified. After some experimentation, the process known as Raman spectroscopy was chosen. Laser light is shown onto molecules, causing vibrations, and, as it turns out, the vibrations differ between healthy and malignant cells and *this difference can be heard.* We can listen to cancer. And, with development, this process could, ideally, be set up in a physician's office. Another anecdote: I described this process for the first time in Toronto, Canada, and in the audience there was a postdoctoral student who was, in fact, undergoing training in listening to cancer. We had a serious discussion after. There are two companies in Canada developing the process and, as the *Scientific American* article also notes, there is some concern about how deeply entrenched the visualist approach may be, which in turn may complicate the introduction of a new process. "Although sonification offers advantages over visual display, [science sonifiers] face a major hurdle: simply getting researchers to try this

10. Cowen, 47.

new way of exploring data."[11] For, in spite of studies that show that training listeners to discriminate healthy from malignant cells can be done to achieve 95 percent accuracy in an hour's intense training, the long experience of visualism remains.[12]

So far a sonification diagnosis has been limited to specific types of cancer, such as cervical and prostate, but detection of much harder to find bloodstream cancers is being investigated. It is my hope that science and art practices could, and should, enrich each other and, in keeping with the whole-body emphasis of postphenomenology, equally enrich human experience overall.

I add as a sort of postscript that many of the late-life medical procedures I have experienced were sonograms, particularly used by urologists but also for other arterial and esophageal areas.

11. Cowen, 47.

12. See my *Expanding Hermeneutics: Visualizing Science* (Evanston, Ill.: Northwestern University Press, 1998).

Aging: I Don't Want to Be a Cyborg, I and II

ALTHOUGH DONNA HARAWAY did not invent the term cyborg, she did help make it a term of popular culture. Her cyborg is a *hybrid* that can include human, animal, and machinic or technological parts. Haraway's cyborg is simply a mixture. This, in turn, may combine with various human fantasies, including what I call *technofantasies.* Haraway herself was also well aware that the cyborg could stimulate imaginative fantasies. In popular culture, these fantasies include utopic-bionic science fiction variants that in film and television include all sorts of prostheses that make mere human limbs, organs, and the like superpowerful and better than the originals. *Terminator, RoboCop,* and *The Six Million Dollar Man / The Bionic Woman,* have all had their plays upon this theme. And each is a variant of what I claimed was a deep-rooted technofantasy in *Technology and the Lifeworld* (1990):

> There is a . . . deeper desire which can arise from the experience of embodiment relations. It is the doubled desire that, on the one side, is a wish for *total transparency,* total embodiment, for the technology to truly "become me." Were this possible, it would be equivalent to there being no technology, for total transparency would *be* my body and senses; I desire the face-to-face that I would experience without the technology. But that is only one

> side of the desire. The other side is the desire to have the power, the transformation that the technology makes available. Only by using the technology is my bodily power enhanced and magnified by speed, through distance, or by any of the other ways in which technologies change my capacities. These capacities are always *different* from my naked capacities. The desire is, at best, contradictory. I want the transformation that the technology allows, but I want it in such a way that I am basically unaware of its presence. I want it in such a way that it becomes me. Such a desire secretly *rejects* what technologies are and overlooks the transformational effects which are necessarily tied to human–technology relations.[1]

In that earlier context I was describing what I have called *embodiment relations,* which were experienced uses of technologies that remain detachable but that in use are quasi-transparent and not literally taken into or inside my body. Yet, I will maintain that the desire remains applicable to *cyborg technologies.* To make the case, I will need to account for several historical as well as contemporary variations upon technologies used both in detachable and nondetachable, internalized forms.

One of the oldest such anticipations of cyborgization are *prosthetic devices.* False teeth, peg legs, arm hooks, and various other devices to replace lost teeth, limbs, and such are very ancient. These prosthetic devices, substitutes for lost body parts, remain detachable and thus fall under the earlier descriptions I have made concerning embodiment relations—one experiences one's surroundings through the quasi-transparency of such devices but always with a detectable difference that magnifies some and reduces other features of one's experienced environment. The peg leg can never "feel" the hot, sunbaked sidewalk that the bare foot would feel—but slid along the rough texture, one might even

1. Don Ihde, *Technology and the Lifeworld: From Garden to Earth* (Bloomington: Indiana University Press, 1990), 75.

better than with toes "feel" the rough textural features of the surface. Nor would I expect that the users of such ancient devices could easily fall into the slippery slope and utopian fantasies that so frequently dominate science fiction and virtual-reality hype contexts and that describe these devices as "better" than the lost body part—or am I wrong? *Pinocchio,* after all, capitalizes on a fantasy of a dummy-become-alive! The desire remains, but the device remains "dumb." No one in a right mind would likely seek an amputation that would replace one's healthy limb with a wooden one. But most people would choose to have a prosthesis once the limb is gone in order to restore some semblance of motility and capacity. The proto-cyborg is thus a compromise.

Vivian Sobchack, a phenomenologically trained scholar with a high-tech prosthesis, brings us up to date in a specific response to the quotation above:

> Obviously, transparency is what I wish—and strive—for in relation to my prosthetic leg. I want to embody it subjectively. I do not want to regard it as an object or to think *about* it as I use it to walk. Indeed, in learning to use the prosthesis, I found that *looking objectively* at my leg in the mirror as an exteriorized thing—a piece of technology—to be thought about and manipulated did not help me to improve my balance and gait so much as did *subjectively feeling* through all of my body the weight and rhythm of the leg in a gestalt of intentional motor activity. So, of course, I want the leg to become totally transparent. However, the desired transparency here involves my incorporation of the prosthetic—and not the prosthetic's incorporation of me. This is to say that although my enabling technology is made of titanium and fiberglass, I do not really or literally perceive myself as a hard body—even after a good workout at the gym, when, in fact, it is my union with weight machines (not my prosthetic leg) that momentarily reifies that metaphor.[2]

2. Vivian Sobchack, *Carnal Thoughts: Embodiment and Moving Image Culture* (Oakland: University of California Press, 2004), 172.

This is a good description of, precisely, an embodiment relation, of *quasi*-transparency. Sobchack, however, is not tempted by the slippery-slope utopic slide:

> Nor do I think that because my prosthetic will, in all likelihood, outlast me, it confers on me invincibility or immortality. Technologically enabled in the most intimate way, I am, nonetheless, not a cyborg. Unlike Baudrillard, I have not forgotten the limitations and finitude and naked capacities of my flesh—nor, more important, do I desire to disavow or escape them.[3]

Even more high-tech prostheses built with springs for below-the-knee amputations have allowed highly motivated and skilled athletic persons to actually achieve high running speeds, such as those demonstrated by Jami Goldman, amputee sprinter. Perhaps the most famous model-athlete is Aimee Mullins who, with her spring legs, played Cheetah Woman in Matthew Barney's *Cremaster 3* movie series. Mullins turns in highly respectable records running but also alternates her spring legs with other prosthetic legs, including a pair of glass limbs, for other purposes. She has had prostheses since learning to walk because she was born with fibular hemimelia (born without fibula bones) and underwent amputation at age one. Growing up with prostheses is probably as close as one can get to minimal quasi-transparency. Prostheses with spring components have become more common with the large numbers of Iraq War amputees as well, yet all remain within the noted degrees of limitation cited for detachable devices.

Before leaving limb prostheses, I should mention their internal and permanent counterparts—knee, hip joint, and other implants—substituting for bone and cartilage damage. Stainless steel and Teflon restore the motility lost to damaged joints or ar-

3. Sobchack, 172.

thritic deterioration to a degree. But while the metal and plastic implants, insofar as the materials are concerned, might "outlive" the patient, in practice the stress and strain usually call for replacement on a seven- to ten-year basis. (Newer prostheses are said to last for up to thirty years.) And since more bone has to be removed for situating each new replacement, diminishing numbers of such replacements are possible for finite human lifespans. (All prostheses, older or newer, have finite shelf lives.) Thus one must hope for late life rather than midlife cyborg parts!

Much more common, but still detachable, are *sensory* prostheses such as optics and audio technologies (eyeglasses, contact lenses, hearing aids). Again, there has been a long history to such body-related, sensory-correcting technologies. Eyeglasses were already common in Europe by the thirteenth century, and one noted social effect was to prolong careers for scribes and accountants beyond the age when one normally needed reading glasses, thus closing off what had been in pre-eyeglass eras jobs for younger scribes and accountants. One could continue to read into old age. Contemporary contact lenses, while still detachable, are much closer to quasi-permanent cyborg capacity. And while I have no experience of contacts, my family members do, and it is clear to me that the occasional dust or eyelash occurrence, torn lenses, and the like retain the sense of compromise I suggest belongs to these technologies in use. In my own case, I did not need reading glasses until age fifty-eight when the *New York Times* and telephone directories became unreadable. These I still must use for fine print, but my distance sight remains such that no other optics are needed. (As noted later, cataract surgery has allowed me to abandon reading glasses!)

In the case of loss of hearing, hearing horns have been depicted in treatises on the senses for several centuries. But hearing horns are simply amplification devices, and in most loss of hearing, more is needed than mere amplification. With normal aging, most people began to lose certain—usually high—frequencies. These cannot be restored with amplification. In my late sixties I, too, began to notice some hearing problems and noted difficulty hearing questions from the back of lecture halls and found cocktail party conversation hard to manage. Taking a frequency range test during a conference in Boston at the Museum of Science, I found that my hearing was considerably short of the 20,000 cycles per second younger, better hearing could detect. Today, I wear state-of-the-art digital hearing aids, and I could echo Sobchack's desire for transparency of hearing, which remains much more difficult to attain with hearing technologies than in seeing technologies. I have described in detail the comparative embodiment processes in the second edition of *Listening and Voice: Phenomenologies of Sound* (2007). Two of the points made in such a phenomenology are the far greater difficulty learning to rehear and the far greater complexity in the technologies between optics and audio devices. For example, whenever an optical prescription is changed, the user experiences very subtle changes in motility and spatiality when walking for example, but it is but a matter of days at most before the quasi-transparency is "fully embodied" and no longer occupies any significant role in daily life. Contrarily, the auditory embodiment in using hearing devices is much slower and more difficult to attain, so much so that many individuals give up or reject using hearing aids entirely—a fact well known to audiologists. And, even with the best and most expensive devices, feedback at certain frequencies, the impossibility to match normal hearing's capacity to sensorially inhibit background sounds (as in a cocktail party situation), and full auditory transparency remain noticeable long after one becomes accommodated to the devices. Even one's musical memory reminds one that music no longer "sounds the same." Frequencies lost remain lost, but high-tech digital devices can partially compensate for the loss of consonant sounds—which are more often lost to those with hearing disabilities compared to vowel sounds, which are more easily amplified—thus making speech more available than would be the case without the prostheses. They remain worthwhile trade-offs, but they also remain short of full transparency in user experience.

Closer to a cyborg body notion, implantable devices display somewhat different characteristics, and it is at this point that I shall begin with some self-reference related to "my case" and my reluctance for cyborg status:

- It began quite a long time ago with dental technologies—tooth crowns—of which I now have five, plus a root canal (living tooth nerve replaced with a filler). The first broken tooth was due to a piece of sand in a Maine mussel. The silicone of the sand proved harder than my tooth enamel, and so while still a graduate student, I got my first taste(!) of cyborghood. Interestingly, I can still detect the differences experienced with crowns. They "feel" different to the tongue: they lack the striated texture of an original tooth, they can cause chewing gum to stick, and the shape is never quite the same as the original tooth. While feeling-through-the-tooth as with eating has become indiscernible from the original teeth, feeling the crowned tooth with my tongue retains a discernibly different feel. Thus, though permanent, the crown retains a marginal self-difference. But, given the choice of either a missing tooth or an aesthetically odd-shaped one, I would willingly choose the cyborg crown.

- Making a vast leap into animal or other human *transplants* (heart, heart valves, lungs, kidneys, etc.), I feel fortunate enough to have so far avoided this degree of cyborghood [see follow-up below]. Yet, one interesting study just now getting underway at the University of Toronto, in the Health Care Professional Training group, has noted anecdotal evidence from heart transplant patients that indicates that such a major transplant trauma often leads to an *experienced personality change*. Patients claim to feel as if they were "another person" or are incorporating another person after the procedure. A carefully designed study is planned to investigate this phenomenon.

- An electronic transplant is yet another variant—pacemakers and defibrillators, for example. Pacemakers are devices either implanted under the skin or worn on belts with only wires implanted, which electronically pace heartbeats for patients whose beats are irregular or too slow. Today, some are even equipped with radios to signal to a central medical facility should trouble occur. As with many contemporary procedures, the implantation can be performed and the patient released the following day. Most felt responses are not so much related to the experience of a more regular or faster heartbeat—although this is experienced—as much as hoped for more indirect results, such as greater energy levels and less fatigue with daily activity. Some minimal training for what to be aware of is taken before hospital release. Patients describe less light-headedness, incremental energy improvement, and less "stuffiness" in their chests. Most agree that the implantation was positive enough to confirm a right choice for the procedure.

- This brings me to my own next stage cyborg addition—a medicated *stent* in one of my heart arteries. I had long delayed having a recommended colonoscopy but finally agreed to undergo the procedure. My physician, upon doing a pre-procedure check, detected—with a stethoscope—a slight heart murmur. It is worth noting that the stethoscope, although now an ancient device mostly used as a symbol of practicing medicine as per television commercials, in skilled practice can reveal very nuanced internal phenomena. Auscultation, or listening-to-interiors, was a favored diagnostic art at the end of the nineteenth and on into the twentieth century. But, as articles concerning medical education have pointed out, it is for the most part a lost skill amongst many contemporary physicians. So then, following the dominant *visualist* and *instrumental* practices of contemporary medicine, a series of tests was ordered: EKG, or the electrical graphing of heart motions; echocardiogram, a multimedia visualization imaging that produces dynamic images of heart motion with added graphing of the beats; *and an audio counterpart as well.* I have been researching imaging technologies for more than a decade now, and when I inform my physicians of my interest, I almost always get a detailed and interested response with demonstrations of how the imaging is interpreted. I also collect copies of the actual imaging performed! More then followed, with stress tests and before-and-after imaging and finally a recommendation to undergo an *angiogram.* This procedure entails a small incision in one's femoral artery (in one's leg alongside the groin), insertion of a catheter with fiber-optic lights and internal instruments for further interventions, and movement of the catheter Nintendo surgery–style (guided by imaging on a screen and manipulated by a set of "joystick-like" controls) up into the arteries of the heart itself. A radioactive dye is used to show any obstructions, which when found may be forced open by balloon angioplasty, that is, the obstruction is shoved aside by an inflated mini-balloon or, if necessary, the insertion of a stent. The stent is a small wire mesh tube, often impregnated with a slow-acting drug to prevent clotting. And, in my case, one balloon procedure was performed and one stent inserted. As with pacemaker implantation, I was released the next day and while rest was prescribed for a limited period of time, I was soon back at work.

- A phenomenology of this event is relatively simple. I was kept awake so as to move as directed. I could see the screen, which was also moved repeatedly to the advantage of the operating surgeon, and feel the warming infusion of the dye but very little pain or sensation of the catheter in motion. It seemed a quite minor interruption of life, and the result was, indeed, an incremental improvement in energy levels following. I, again, have the entire procedure recorded on a DVD that I now use as a demonstration of a medical imaging technology on the road. Admittedly, it was somewhat disturbing to find—only a few weeks after my operation—

> that drug-embedded stents were found *not to be as dramatically effective as first thought*. Some patients were found to develop clots long—several years—after the operation, and thus while stenting remains more common than arterial bypass surgery, its effects are more parallel to bypass than previously thought. Nevertheless, had I to do it over, I would still choose the stent rather than the much more intrusive bypass open-heart surgery! Interestingly, I have no direct bodily awareness of the stent at all—unlike my tooth crowns, it remains totally "invisible," although the indirect awareness of energy levels may be noted. I am thus a limited or partial *cyborg*.

By now it should be obvious that the gradual accumulation of human–technology hybridization, or the cyborg process, often relates to effects of contemporary *aging*. One's eyes gradually lose the flexibility of younger pupils, lenses, and orbs; the cilia in one's inner ears deteriorate with age (in my case the "boilermaker" sounds of a John Deere tractor may have been my equivalent of the rock band sounds of my children's generation); and plaque in one's arteries tends to build up with age. Thus, as illustrated above, cyborg strategies are often technological attempts to thwart even more severe effects of aging. I have not dealt with animal transplant strategies as fully here, but such transplantation strategies also have their associated problems, such as transplant rejection. Furthermore, the cyborg strategies show that such delaying tactics remain trade-offs, compromises. It is better to have a pacemaker than to have life-threatening arrhythmia; it is better to be able to walk with either a steel-Teflon implant or a prosthesis than not to walk at all; it is better to have digital hearing aids that allow seminar participation and exchange than not to be able to hear speech sufficiently to understand. Yet all these trade-offs fall far short of the bionic technofantasies so often projected in popular culture or post- and transhumanism.

What, then, motivates the continuance of technofantasies, the unrealistic imaginations of utopic cyborg solutions to our existential woes? At the popular level, the sheer entertainment value mixed with the contradictory desires we have concerning our technologies may provide some of the answer. "Explosion movies" remain popular—the quasi- or even superpowers of a Terminator or a RoboCop indulge wish fulfillments and even revenge fantasies. But they also reverberate with our secret desires to be bionic technology plus human, equaling superpowerfulness. Such fantasies are, of course, ancient, but in other cultures and times they did not always take technological form. To be godlike, to have animal powers, and to be more-than-finite human are desires reflected in literatures and magical practices on a wide scale. Ours, reflecting the highly saturated technological texture of contemporary life, more often take technological or cyborg forms. Not satisfied with quasi-transparency, we seek the organic cyborg solution that does not happen in actuality or mundane life.

There may also exist as a motivation a social or even industrial desire to keep utopic hopes stimulated as possible sources for technological development itself. Could, for example, technoscience development remain well funded were all utopic fantasies to disappear? I suspect that it is not accidental that Stephen Hawking urges humankind to think about escape to other planets. He is convinced that we will end up destroying ourselves and thus need to plan for extrasolar life. Yet, as a physicist, he must be aware that so far the only possible life-supporting planet recently discovered lies some twenty light-years away. No existent technology can even approach the speed of light, and thus such a journey seems close to yet another technofantasy. I am suggesting that there may be a subterranean link between the wildest science fiction fantasies and our current technological culture. Do the desires, dreams, and fantasies indirectly support financing of research and development?

Or, is it that the deepest desires and fantasies are simply our wishes to avoid mortality and contingency? And if so, is my own reluctance to cyborghood part of this same phenomenon? Should I not, contrarily, argue that precisely to accept finitude and contingency applies equally to cyborg technologies? That is, to accept the cyborg destiny is also to accept the trade-off compromise that all our actual technologies display, and this, in turn, is existentially tied to the human process of aging.

Epilogue: Aging Cyborg II, More Cyborg than Ever—Open-Heart Surgery

My stent was implanted in 2006, just before I completed this article on aging for its first appearance. It is now 2008, and I have become more cyborg than ever. Once one enters today's high tech medical system, the monitoring, even surveillance, process continues. By the end of 2007 I found myself often feeling highly fatigued and short of breath. My cardiologist ordered stress tests and more echocardiograms, and later another angiogram. All revealed more obstruction to my heart arteries—again, odds are that approximately half the arteries cleared by angioplasty can again close within six months to a year—and imaging showed a severe "regurgitation" with my mitral valve. My cardiologist recommended that I consider surgery and recommended a mitral valve specialist. So, by watching my imaging CD, turning to research on the internet, and above all investigating the specialist who was recommended by my doctor as the "rock star" of mitral valve surgery, I ended up scheduling a consultation with David Adams at Mount Sinai Hospital's famed cardiovascular clinic. From his many entries on the internet, which included video clips from actual operations, reprints of his articles, but above all his fervent philosophy that argued that *repair*—retaining one's own flesh rather than *replacing* it with pig or cow parts or even a metal valve—was far superior and, where possible, yielded better results. The only machinic artifact to be involved was a plastic ring that would "reseat" the valve (a description that reminded me of automobile repair). Here my desire to be less rather than more cyborg kicked in.

Upon arrival at David Adam's office, I felt almost as I did when I played with the cybertoys in Umea University's Humlab a couple of years before. There were plenty of cybertoys here as well. Adams, along with several of his team members, awaited in a large office with a contemporary array of visual display screens. Upon these, of which there were many in the multiscreen equivalent of a newsroom, he played different takes on my latest echocardiogram and TEE imaging, all in dynamic, colored imaging. With his cursor, he showed me that two of the cords attaching the valve to my heart were broken. He showed the backward flow of the regurgitation and many other clearly imaged features. He confidently noted that this made me a qualified patient and that *for him* this would be a *routine* surgery (not for me since it would be my second surgery ever). We concluded with a good discussion of the progress in imaging that characterized such high tech surgery in the twenty-first century. As for risk, mortality was significantly below 1 percent, but with full success, quality of life could be considerably better than before or without surgery.

A bit over a month later, I entered the hospital, somewhat shocked to see that the protocol called for a *triple bypass* in addition to the mitral valve repair! I was to receive a major reconstruction. Here any first-person phenomenology became disrupted—I had no experience from the time going into surgery until waking up in the intensive care unit, now hooked up to a large array of machines and monitors, including an external pacemaker, electrodes from my chest, a quite large leg drain from the area where my vein had been "harvested" for the bypasses, and another drain from my chest. I shall cut short more details, but by the end of a week I was ready to be discharged and returned to Long Island. I was optimistic; I had had none of the frequent complications such as stoke, dizzy and disorienting experiences, lung congestion, or severe pain. I had only a bit of atrial defibrillation said to be common right after such a surgery. And I was so relieved to have my electrodes removed, the pacemaker taken away, and to head home to my first decent meal in a week. A follow-up with my cardiologist on Long Island and another echocardiogram showed that the repair had been successful and my heart functions were excellent.

However, just in case, my cardiologist recommended I wear an "event monitor" to record any unusual rhythms from my heart. Unfortunately for me, the arrhythmia had not disappeared, and a search of earlier EKGs indicated that there had been signs of preexisting irregularities. Note here again the role of consistent monitoring and surveillance. I had escaped major cyborgization with repair and no installed defibrillator or pacemaker only to wonder if I was fated to have one of these installed later—I definitely did not want to be the first philosopher Dick Cheney! I felt lucky, then, when first my cardiologist, Michael Matilsky, and later Jeffrey Mattos confirmed the earlier opinions of experts in Mount Sinai Hospital, all concurring that a pacemaker seemed unnecessary, recommending instead a heartbeat regularizing medication. There was a catch—one must initiate the prescription only while in a forty-eight-hour, full monitoring situation *in a hospital*. There was a risk of a side effect that occurs in less than 1 percent of patients: a worsening of the symptom or even a heart attack. Back to both monitoring and risk-taking. Fortunately, for me, nothing happened, and my heart is currently very regular.

And now, the good news. Although it has only been three

months since my surgery, my old energies have returned; I am back to my strenuous travel schedule and in fact feel better than I have for several years—this had been the good prognosis from such surgeries, which could include a prolonged lifespan in an era in which lifespans are considerably longer than a century ago. Am I more cyborgian than before? Yes, and again related to the aging process, but also minimally more cyborg than could have been the case.

Here, again, one discerns the role of high-tech processes ranging from the improved imaging devices used for the monitoring and diagnosis to the chemical technologies of medications to the complex medical care system but also including all the risks, assessments thereon, and trade-offs entailed with all technologies. In this case, I do not feel so bad about this particular increase in my cyborg identity. Better to be partially cyborg than dead.

Aging Cyborg, III, IV, V, VI, and VII

I CONTINUE my autobiographical-postphenomenological narrative about the experience of an aging cyborg. I left this narrative in the last chapter with a section on open-heart surgery in 2008. At this time I was still a full-time faculty member at Stony Brook. By 2010, my knees begin to pain me when walking or cross-country skiing, a favorite "focal activity" I shared with Albert Borgmann, another philosopher of technology. X-rays showed the signs of aging knees. Thus begins Aging Cyborg III. (Note that X-rays were discovered by Roentgen, who made the first imaging devices in 1895. See a wonderful science history by Bettyann Holitzmann Kevles, *Naked to the Bone: Medical Imaging in the Twentieth Century,* published 1997.) My X-rays showed that both knees had lost half their cartilage so that one half each was now bone-on-bone, thus the pain, but half of each knee retained cartilage. In response, my orthopedist recommended unicompartmental implants (i.e., half joint implants). Later I learned a bit about the economic policy regarding such implants. There had been an overproduction of the devices, which up to my time had only been recommended for younger patients. To lower the surplus, the medical community began recommending use for older patients. There are always economic politics behind protocols! I did not know this when I concurred, so at my Stony Brook University Hospital, in a single day, I underwent—again with local anesthesia—a double implant procedure. Knee implants were thought to be the second most successful surgery of the late twentieth century, with several hundred thousand performed per year in the United States. I was then seventy-six years old, and the older one is, the greater the risk from any general anesthesia. This is why in many of the incidents I cite that only local anesthesia is used. Forgive the gory description that was totally painless: I am lying on a bed, a sheet strung across my stomach to screen sight of my knees. I nevertheless hear and smell the grinding of bone but feel nothing. The operation itself is not prolonged, and I am sent to recovery with my new unicompartmental stainless steel and plastic implants. After physical therapy, although there is some residual stiffness, no pain and full walkability soon return. By wintertime, however, I note cross-country skiing is not what it once was. Most obvious is a loss of finer balance, so now trails need to be restricted to wimpy, flat terrain. All is well until the summer of 2013 when suddenly my right knee begins to pain me. My family and I are, as usual, spending our summer in our mountain retreat in Vermont. Our local clinic's doctor diagnoses an infection three years after my first implant surgery—he and the expert orthopedists at nearby Dartmouth Clinic in New Hampshire panic. A culture is called for because if the bacterial infection is aggressive, the implant must be immediately removed. If the bacterium is active, it could spread and endanger my very life. It takes several days for an analysis to show a result and at first Linda and I think I may have to undergo surgery and end up with long recovery time—minimally five months—in Vermont. Fortunately, the diagnosis reveals that the bacterium is "indolent," so we can delay the surgery for a return to the Hospital for Special Surgery in Manhattan, now end of summer 2013. Of course the replacement process will be slow, and all my scheduled activities for the entire fall term are cancelled. It will turn into a full fall delay before travel can resume. First there is a five-week removal and sterilization process of the unicompartmental prosthesis; once the old prosthesis is removed and my wound is certified sterile, a new full knee implant may be inserted. Rock star surgeons, however, are hard to schedule. Each process occupies a number of weeks. The temporary implant is mostly composed of a type of concrete, which, despite being excellent for physical therapy, is clumsy and lacking flexibility. The ultimate prosthesis is again stainless steel and plastic and, while it causes some residual stiffness, is painless. While this ends my many decades of cross-country skiing, it allows me to walk well if somewhat slower than prior to the implants. I, in the midst of this, also continue to philosophize, and once I am on the road again, one event is a conference in Arizona with a group of transhumanists who have a utopian-bionic view of human enhancement. As mentioned earlier, I often call this a "technofantasy," and when it comes to any kind of internal technologies, in addition to the experimental compromises, there are also technobiological ones. In the case of knees, any time metal or plastic is placed, the body reacts with a *biofilm* that adheres to the flesh/implant surface. So long as such a film is "friendly" or sterile it is harmless, but it can also cause infections. These are unpredictable and thus are or can be toxic. They constitute a biotechnic ambiguity and a barrier to any bionic technofantasy. So, once again, being an aging cyborg is a positive trade-off from immobility or a clumsy electric scooter or wheelchair but not an immortal, bionic enhancement. Being a cyborg is a compromise and definitely not a bionic miracle, which is a mere technofantasy. Technologies, metaphorically "mortal," have shelf lives.

Aging Cyborg IV

Yet aging inevitably continues, and next comes a surprise diagnosis with a medical technics process, this time in 2014. Long ago I used to be a somewhat skinny guy. By midlife I began to have a mild "potbelly" as I gained weight with age. Now five years past my heart surgery, I began to again have breathing problems and some symptoms similar to an acid reflux, so once again a series of imaging technologies were suggested: I underwent CT scans, an X-ray, a barium GI track endoscopy (using a tube device equipped with a light and scope), and another imaging of my entire esophageal tract. But here I need to insert other autobiographical elements. The reason so much of the imaging was directed at my GI tract relates to the then recent death of my younger brother, Jon. He died of esophageal cancer in fall 2014 at age seventy-seven and had, as part of that symptomatology, severe gastric acid reflux. I worried after taking several trips to see him in his last year and decided I had better check my own GI tract to make sure I didn't have anything similar. I didn't: all the imaging came back showing no cancer, no polyps or lesions, nor any acid reflux. However, my very sharp gastroenterologist did detect a problem: gastric volvulus or "twisted stomach." After several consultations it turned out that after heart surgery my diaphragm had weakened, and part of my stomach had actually crept up into my chest cavity, a paraesophageal hernia. So as with my "rock star" heart surgeon, I found a similar rock star GI surgeon who used a laparoscopic process that left five "Machine Gun Kelly" scars on my stomach. He successfully moved my twisted stomach back into its proper place, and after recovery the only bad result was an even bigger and lower potbelly.

Aging Cyborg VI

Here I want to insert a variant on my autobiographical-postphenomenological narrative in the form of a noninvasive and positive result event. I call it my eighty-second year scare. I have already included the events around my younger brother's death at seventy-seven. As it turns out, both my mother and father also died at seventy-seven. My father died of the ultimate effects of prostate cancer, that bane of males. He was a Kansas farmer, and after his prostatectomy at age sixty-three, I visited him and once went with him to consult with his urological surgeon. I was struck, as I waiting in the office, to see a room full of aging farmers—all with prostate cancer. So, when we finally entered the office for the consultation, I asked, "How many of these men with prostate cancer are farmers who use mercury gas treatments for stored wheat?" The doctor answered, "All of them." Remembering how hot granaries become in a Kansas summer, I knew that my father never used the recommended protective masks for such grain treatment and suspected the same of the others in the waiting room. Mercury gas is a recognized carcinogen.

The result for me was to have all my urologists from middle age on worry about the genetic likelihood of prostate cancer for me. My university urologist was, in fact, so insistent upon recommended biopsies that over the years I succumbed to five of these, all involving sonograms and all of which returned negative. Later, my brother also underwent a prostatectomy, raising further my earlier urologists' genetic concerns. Upon retirement, a new Manhattan urologist worried about my PSA levels (a notoriously problematic and often undependable blood test) recommended that I undergo a new urological MRI exam (no biopsy). After an hour inside the MRI machine—very noisy—the test came back showing a perfectly normal prostate. In this case a newer technology was liberating. My annual results remain normal.

Aging Cyborg VII

Aging continues, and I return to perceptual prostheses and very minor and common procedures. I have always had excellent vision although in my sixties I did have to resort to reading glasses—and at about the same time to hearing aids. Again, some autobiography: I begin with listening prostheses, hearing aids. It is common knowledge that both eyesight and hearing decline in aging. It is less clear when and how one becomes aware of such loss. By middle age it is common for an aging person to be aware of loud ambient noise situations in which it becomes more difficult to hear conversations. By my sixties I was aware of this loss, but as the author of two books on sound and acoustics and with considerable background knowledge of hearing capacities—for example knowing that adolescents can often hear from .20 to 20,000 hertz—I knew that as one ages there is a gradual loss of higher frequencies. Once at a conference in Boston, while in my sixties, my son, Mark, and my wife, Linda, urged me to join them in a visit to the Museum of Science, Boston. One of the exhibits had a sound device where one could put on earphones and turn a dial to test one's hearing perception, a diagnostic technology. I did my test and discovered that my highest frequency range was a little over 11,000 approximate hertz. I panicked—could it be that bad? So as soon as I returned to Long Island, I did a search in my *Macropædia* only to discover that my limit was about normal for my age. After that I began to research hearing aids. Once I took an audiology exam that focuses upon language perception with the advice to begin with two, not one, aids (in order to retain the perception of direction and distance). I ended up with my first digital aids. During this process, I learned how focused upon language perception such devices are. For example, vowels are easy to amplify and need little manipulation. They have a longer sound *durée* than consonants, which are shorter and call for digital manipulation. It is the lack of consonant perception that leads to lessened language comprehension (combined with loss of high frequencies, which sometimes makes females harder to hear than males!). But here technological advance meets the low-grade American health-care system. Improvements in acoustic technologies outspeed the four-year limits of hearing aid insurance! Thus as I aged, and particularly after retirement, I had taken to two-year hearing aid purchases (at a cost, of course, since my insurance only allows a four-year subsidy). My latest device purchase, in a move from a German to a Danish design, was made in 2017. As it turns out, my right ear is better than my left, which has deteriorated to the point that my left aid is not actually a hearing aid but a broadcaster which takes all sounds from my left side and sends them to the complex digital hearing aid on my right side. This, in my estimation, is better than a cochlear implant and retains a natural sound. It is a significant improvement upon the previous devices.

Visual technologies, always simpler and more attended to than acoustic technologies, rely upon optics and optical theory, older and simpler than sonic technics. These also entail the experiences of aging. As noted, I had always had excellent vision—interrupted for a short time by two accidental corneal scratches, both occurring during a Vermont summer. My middle daughter accidentally scratched one cornea with her finger. Later that same summer, a pestiferous porcupine began to chew on the foundation log of our then three-room log cabin. I chased it off after dark but ended up scratching my second cornea by running into a tree branch. Those who have suffered corneal scratches know that these are painful and badly distort vision. So, for a time I had to wear prescription glasses.

Fortunately, my vision returned and the only leftover was another period of needing reading glasses.

Upon retirement, the move to Manhattan called for a new ophthalmologist, who discovered very small cataracts in each eye. Even so, I could still pass unaided reading tests so as not to need glasses for driving. But as is normal, the cataracts increased in size until by 2017 a cataract lens procedure was recommended. I opted for a multifocal lens and a new procedure, femtophotography, and scheduled the femtosecond laser process with two weeks between the two eyes. Femtophotograpy is one trillion times faster than stroboscopic photography and can image a single photon in motion. A similar leap in precision goes with this new cataract process. Each procedure lasted about fifteen minutes. As I have noted elsewhere, nanoscale technics are the mark of twenty-first-century technologies. The leap in precision is startling.

While two weeks lapsed between eye processes, in the interim I could blink eyes and see the difference between the cataract and non-cataract eyes: the cataract eye was considerable darker and differently colored than the new lensed eye. Once both were completed, my vision was so spectacular that I could read the very smallest print lines on eye charts, and the hardest adjustment was to no longer need reading glasses. But the acid test is yet to come. Sandra Harding, feminist philosopher of science, claims in *Is Science Multicultural?* (1998) that Dogon people in Africa can, with the naked eye, see the larger satellites of Jupiter as protuberances. I discovered some years ago that I, too, could do this and after sighting this phenomenon on the way home from a concert in Vermont, checked with my telescope to confirm my angles were correct—they were. But so far all our trips to retry this have had cloudy skies. I intend to try again this summer. In short, this is the medical technics that actually comes close to *human-enhanced vision*. But as a prosthesis, it is obviously a permanently installed mediating technology.

This now brings a decade of my life, age seventy-four to eighty-four, to an end. I have recited seven aging cyborg events, but one was simply a diagnosis that turned out negative. These events translate into every other year occasions between which I, although slowed down from preretirement, remain quite active. If anything, retiring to Manhattan means at the least that the city is full of excellent medical facilities, doctors, and hospitals. Everything is nearby and encourages us never to delay seeking advice and diagnosis. In my case, skipping ahead to the last chapter, my aim is to keep going, hoping to equal the longevity of my older peers.

Cyborgian Postscript

I don't want to allow a foreshortened impression that of the seven medical events I describe over these ten years, all are indicative of a deteriorating aging process. If one takes a composite of these events, they are very uneven. Open-heart surgery is obviously a traumatic event—although recovery time in my case was very fast. Here the social phenomenon of what I am calling the rock star surgeon is important. Manhattan is concentrated with many of the world's best hospitals that in turn accumulate rock star surgeons—specialized doctors who do *only,* or nearly only, one thing hundreds or approaching thousands of times per year. This was the case with my heart, gastroenterologist, cataract, and knee replacement surgeons. But whereas heart surgery is traumatic, cataract surgery is not. Some 3.6 million per year are performed in the United States alone. Cataract surgery, now a laser procedure, takes fifteen minutes per eye, and this implanted visual prosthesis results in improved vision by day three. Knee implant surgery is followed by weeks of physical therapy, but within months motility can be cane free; this, too, is an implanted prosthesis. Hearing aids are removable and are familiarly embodied within a few days. Thus, while seven events in ten years may seem like a lot, recovery and embodiment learning time has been relatively fast in each case.

From Embodiment Skills in Computer Games to Nintendo Surgery

AS NOTED IN MY EARLIER LLLB (late life little book), *Ironic Technics,* much of today's surgery is often dubbed "Nintendo surgery" with the idea that the new skills in this microsurgery relate to the eye–hand skills common to playing computer games. I have often pointed out that when the screen games that dominated computer games first appeared to addict so many young players, much publicity worried about a downgrading of more active, particularly outdoor, play by adolescents, but in retrospect, we can see that this practice ended up being a sort of pre-skilling for what is commonly called Nintendo surgery. (What in critical theory is often decried as de-skilling in this context can also be pre-skilling.) I could not but experience this similarity when I had my first angioplasty in 2006. My surgeon operated a thin tube device that entered my artery from the groin. This laparoscopic probe was outfitted with a light, a balloon device, and eventually a drug-saturated stent. There was a multiscreen display that he could view directly but I could also view from a side angle. I had previously experienced a similar observational experience with older-fashioned colonoscopies.

In the colonoscopy case, while awake I could view the multicolor screen that showed the lit-up interior of my intestine; but how could I tell it was mine? This was a first experience of hard-to-identify imaging, a sort of simultaneous first- and third-person feeling of sighting. For a philosopher, the later style of colonoscopy where the patient is anesthetized is a loss. After turning eighty, physical colonoscopies are considered too risky to take at all.

My highly skilled heart surgeon performed the angioplasty while I, with only local anesthetic, could watch the process. Phenomenologically this is an odd experience since it doubles a seeing of oneself both as in an active sight and as a sort of "body-object." Years later, giving a lecture on "My Case," which drew from the various DVDs I have from such procedures at Oxford University, my audience took note of this oddity of a doubled self-observation. Many of the audience members were queasy while others intuited that this simultaneous subjective/objective vision approximates the skilled vision of the surgeon.

Much later, my audiologist who tests for hearing changes referred me to a soon-to-retire acoustic surgeon in the practice. He was retiring early, he told me, because the "young surgeons who use Nintendo surgical skills" were much better than he was, trained to use different technologies. A similar practice pre-skilling occurs with simulation training for pilots and even more for the long-distance spectrum of remote sensing and controls for Mars Explorers and military drones.

Although this may seem somewhat of a detour, I turn to a similar analysis of computer game skills that appeared in *Debugging Game History* (2016). The reader can note the relation between these preparatory skills and their new embodiment in surgery.

Embodiment

This will be a postphenomenological analysis of variations on embodying screen games—video, computer, arcade, and Wii-style bodily play skills. Although gaming with screens, in contrast to pinball and other earlier mechanical arcade games, began mid-twentieth century under the dark designs of AI combined with war gaming, this approach will concentrate upon entertainment games and look at typical variants through the mid-twentieth to twenty-first century. In each variant the player must develop bodily skills, some of which have far-reaching implications with respect to contemporary life practices.

The analysis will be interrelational and parallel to the many styles of such analyses found in science technology studies but will focus upon the postphenomenological interest in bodily perception and action, and thus the player remains one focal interest. But a game presents a "world" in the sense that some scene of action is presented or imaged—for purposes here, usually on a screen or visual display device. The interaction involves bodily perception, movement, and skill development that varies with the multiple "screen worlds." Embodiment in relation to technologies, however, takes different shapes. Indeed, each type of technology calls for different actional skills. Varying degrees of bodily engagement and skill levels ranging from amateur to virtuoso also come into play. In my own case I have done studies in relation to musical instruments ("Postphenomenology: Sounds beyond Sound," forthcoming in Cohussen, Meelberg and Truax, *Routledge Companion to Sounding Art*) and a range of technologies in my *Listening and Voice: Phenomenologies of Sound* (2nd ed., SUNY Press, 2007) chapters 12 to 13 and in my *Experimental Phenomenology: Multistabilities* (2nd ed., SUNY Press, 2012) chapters 12 to 14. Embodiment is a dynamic and complex phenomenon. It entails learning bodily skills and thus is developmental, and different technologies call for different degrees of engagement—take two keyboard examples. Typing or word processing calls for speed skills and eye–hand coordination but does not ordinarily call for more complete bodily engagement. In contrast, a virtuoso piano player engages the whole body, as any prime performance shows. The same spectrum applies to game skills, although the simpler and earlier games remained closer to the word processing example except for speed.

Two-Dimensional Games

The simplest early games were often two-dimensional: *Pong, Tic-tac-toe, Pac-Man,* and *Raster* were paradigmatic. Here the image worlds were abstract figures that moved upon a flat background screen. The figure, a ping-pong ball, an X and 0, or an abstract gobbler, moved on the screen. The player, using keys or a joystick, controlled the motions according to the scheme of the game. Note that this game world is an analog of a long history of *writing technologies.* Cuneiform was an early hard technology writing process employing a hard stylus and clay or pottery as the proto-screen for an inscription and the skilled scribe who made the inscriptions. Scribes had to follow the writing game to be intelligible and thus learned forms of embodiment. Almost as old are soft writing technologies, brushes or quills (add ink or paint), which produce images on usually soft screen analogues—papyrus, parchment, etc. By industrial times a two-handed approach, a typewriter with a keyboard making images upon paper, appeared. But note that the writer takes action through a device to produce an image world on a screen-type tablet. The two-dimensional game worlds remain analogous to this history; their textual counterpart is of course word processing, which produces its text world on the screen and is later transferred to printed form. *On* screen (or tablet or inscription surface), figures are typically located on stable, opaque backgrounds. Here the bodily motions are minimal. Eye–hand motion is focal. The player remains seated in a stationary relation to the game screen, and thus whole-body motion is minimized. Indeed, the earliest forms often involved monochromatic figures against a contrasting monochromatic and opaque background.

Before leaving this style of player-technology-game world, two features are worth noting: first, game capacities, for example the speeds of balls, gobblers, and the like could be accelerated beyond human reaction times. Thus skilled players could improve their relative eye–hand coordination times to a personal maximum. Second, by the reduced but specific eye–hand movements, new skills could emerge with expert players quite different from a general populace. The unpredicted outcome was a pre-skilling of rapid eye–hand motility that eventually became useful in non-game contexts such as laparoscopic surgery. Unintended effects relate to virtually any technology, and they may be destructive, simply unexpected, or, as in this case, surprisingly useful. Eye–hand coordination skills are, from a perspective of partial bodily movement skills, seemingly reduced bodily actions—but they are also more useful for the meticulous movements called for in microsurgery. Anecdotally, my auditory surgeon tells me he gave up actual surgery some years ago since younger—game-skilled—surgeons now perform better.

Two-dimensional games such as those described remained abstract in image and highly reduced from more full human–game interaction. Hybrid attempts to "jazz up" such games sometimes took the shape of adding colors to the earlier monochromatic game worlds and three-dimensional backgrounds (such as clouds or flat buildings). But a larger shift occurred with games taking on a new dimensionality through a different player-screen variant, that of *through-the-screen imaging* spatiality.

Shooters and Simulations

A marked shift in game worlds began once the figures and characters became more than cartoon-like and took on three-dimensional image characteristics. The games selected here are shooter and simulation games, of which there are hundreds. Again, the focus is upon the interrelational player–game world relationship, with special attention to embodiment skills. The first and most dramatic difference is that the game world appears *through* the screen rather than upon it. The screen becomes mostly transparent (back glare or smudging can occur as an irritant, of course), and the images move in a spatiality that, through the screen, is usually called cyberspace. The two variations now noted are those of screen opacity (2-D) and screen transparency (3-D), which are multistable spatialities. Multistability occurs with most human–technology interrelation in that there are multiple potential uses and outcomes with corresponding shifts in spatialities. Such multistabilities are quite dramatic in that what counts as figure and ground change. In *on-screen* spatiality the opaque screen is ground, stable and flat, but with the introduction of even partial three-dimensionality, the ground can become dynamic with figures that are clearly mobile. With Wii-type games, both player and field become dynamic.

Beginning with shooter games, for example *Duke Nukem,* one of the longest running such games, the player "shoots" whatever counts as an enemy. These games also introduce a multistable set of shooter POVs or points of view. Close up or embodied, the player simply sees a weapon immediately in front of him or her. Different weapons may be used, but all are "as if" being held in the player's hand. Or, an avatar variant may be introduced where some figure stands in for the player and holds the weapon and, either walking or riding some vehicle, shoots the enemies. "Piggyback" is the medium close variant. By this I mean that the avatar is immediately in front of the implied player, not "out there" in the field. Or, the Avatar may be farther away, in the field itself, but again the controller fires the weapon of choice. Thus though the screen worlds introduce many more multistabilities than the simpler games, simultaneously one's bodily actions are more complicated. However, the player remains relatively stationary, seated in front of the screen, and bodily action remains mostly eye–hand motility. Note that here attention remains upon embodiment and the development of player-world skills, not upon plots or narratives. In the case of *Duke Nukem,* both fairy tale–like and science fiction–like plots prevail. These games also frequently entail multiple deaths from which the player may restart and the previously noted speed up such that no player may keep up with the attackers. One wonders—no answer attempted here—how does this re-dying and loss to the game process affect the player's psyche?

Turning to simulations, *Flight Simulator* is the game of choice for this second related category. In these games (which could be simulations of cars, motorcycles, or any number of vehicular machines), a variety of planes are offered, for example a Spad, Cessna, F-14, etc., each with programs determining capacities of flight. Note that this set of choices parallels that of weapons in the shooter games. Then, multiple POVs are also variants. One may be seated in the cockpit, the instrument panel and controls immediately before one and the scene outside viewed through the virtual cockpit window. Or, one may simply have the airplane fly at a distance while projecting an avatar pilot inside. In playing the game, it is possible to get lost and to exceed the capacity of the airplane—and thus crash into trees, buildings, and the like.

These games, in a parallel to the Nintendo surgery (the name given to game skilled eye–hand surgery) example above, have had impacts upon new technology uses. For example, highly complex and realistic simulators are used to test and train actual airline pilots. Programs are introduced for emergency situations that are unlikely or improbable to ever actually occur in flights, teaching pilots how to react and deal with such emergencies. Actual pilots report genuine sweating and stress during such training. Or, in the contemporary world with drone warfare, the pilots are again often skilled gamesters now become remote and long-distance; pilots for flight control work a continent away from the actual battlefield (or, for that matter, even longer distances when controlling Martian space vehicles!). In short, the spatiality-transformations possible in games and remote sensing transform capacities of human–technology relations to mega-space.

Such virtual and remote control again shows phenomenologically how geometrical spatiality differs from experienced spatiality. The near-distance of remote control is a type of bodily or bodily-extended experienced embodiment. Far from disembodiment, this is a distinctively embodied experience.

Learning Lab Denmark

LEGO has long been a major toy manufacturer in Denmark. One of its spin-offs was a gaming think tank called Learning Lab Denmark. Participants in the lab came from many disciplines and were frequently assigned experimental tasks that involved imaging possible games. During the 1990s and early 2000s, a major concern was that children were spending very large stretches of time stationary before screens. Indeed in the United States, recent surveys show that college students average some thirty-seven hours per week either with screens or music devices. Danes became concerned that the resultant lack of whole-body activity would lead to obesity, and thus LEGO launched a study to imagine modes of engaging more complete bodily motility in electronic games. (I was a frequent participant for over a decade.) Workshops included analyses by neurologists, gym teachers, physiologists, and philosophers. Simple ideas such as stationary bicycle races tied to simulation games, 3-D ball games, and the like were imagined. What became known as Wii games were invented independently.

In later meetings of the lab some of these games were demonstrated and, indeed, they called for much greater degrees of bodily motion. For example, players had to stand inside a designated space, and various imaged 3-D balls were fired at the players who, in turn, had to hit them back. A certain irony emerged amongst the players: the adults seemed to enter quite fully into the action and used their arms and legs, soccer-like, to return the balls. But the younger children—probably more accustomed to the mini-movements of screen games—quickly realized that a small and bare hand swat was sufficient to return the ball, so they tended to do much more minimal motions. Embodiment here does increase from eye–hand, but only incrementally. Also, if a player steps out of the designated space, he or she effectively leaves the game. What we are seeing here is both a change in relation to limited, partial bodily motions and greater whole-body motions.

During this same time period, virtual reality caves were also popular. Here, either with goggles or with multi-beamed location sensors carefully tracking the player, and again within delimited space, bodily action was required to discover keys for hidden chests or other in-game actions. These were sort of in-game variants of "Dungeons and Dragons" and yet another way of enticing whole-body motion.

Pterosaurs

Started in 2014, there is an elaborate exhibit at the American Museum of Natural History about pterosaurs, the many species of extinct flying reptiles of the age of the dinosaurs. There were roughly 186 genera of pterosaurs, the largest with a thirty-foot wingspan and the smallest the size of a small bird. The exhibit includes two pterosaur games where one can "fly" two different kinds of pterosaurs. The first is in a sea area on a fairly large animal. Similar to Wii gameplay, the player must stand in a specified area and then, with his or her body, mimic flight and pterosaur action. So, by flapping one's arms the pterosaur takes flight, tilting makes it turn, gliding down to catch a fish it splashes into the sea, and again flapping it takes off. (As a player, this is the closest to an experience of embodied flight I have ever had!). The second pterosaur is a forest dweller, much smaller, and the game's task is to catch flying insects. Again, standing in the prescribed position and mimicking flight motions, one must avoid smashing into trees and the like. This game is much harder to embody, and failures are much more likely than in the first. It becomes obvious to the player that considerable learning time would be needed before a high success rate could be attained. It should be noted that the projected 3-D image is not screen-bound but is much closer to a virtual reality projection. Such game worlds remain limited, and if the player either tries to escape the game world or steps out of the designated space, the game crashes. These constraints relate to embodiment since game embodiment *is distinctly different* from ordinary embodiment. Similarly, game embodiment entails what all aesthetic experience calls for: a suspension of disbelief.

Embodiment Trajectories

What this progression shows—from the simplest 2-D games to the most complex, more fully embodied Wii-style games—is that each game world is related to player embodiment skills. And, as noted, each skill context can reenter an ordinary lifeworld with a newly practicable skill, from eye–hand or Nintendo surgery to distant and remote sensing and robotics manipulations. This is a pattern that could be taken to be a game-to-new mode of life. Games in the contemporary world have tended to take up more and more "real" concerns. For example, the Information Technology University in Denmark has some twenty faculty members whose academic roles relate to gaming. Of course, this is a reflection upon what is today a multibillion-dollar industry. Similarly, a robotics company in Nara, Japan, hired two hundred "roboticist PhDs" in 2008, another multibillion-dollar endeavor. Gamely embodiment, however, clearly retains a human–technology interrelation. From this interrelation there can be a flow into new lifeworlds that are limited only by constraints of imagination.

Postphenomenological Postscript: From Macro- to Microtechnics

IN THIS CONCLUSION, I will return to more strictly philosophy of technology themes, and thus I ascend in some degree from my previous, more autobiographical narrative to a look at what I think are emergent patterns relating to more futuristic trajectories. It may seem odd, at first, to look at the closing of the planned U.S. supercollider in 1994 alongside Watson's furtive glance at Rosalind Franklin's X-ray crystallography images of biological structures revealing the double-helix shape of DNA in 1953. Linearly this seems to be a reversed history, but I shall take a philosophy of technology perspective related to active perceptual and imaging activity. As I will show, this is not actually a linear anomaly. Rather, I shall use these examples to show the radically different dimensions of micro- compared to macrotechnics.

I begin with Evelyn Fox Keller's *Making Sense of Life: Explaining Biological Development with Models, Metaphors, and Machines* (2002). This book was used in my technoscience research seminar and later followed with a roast of Fox Keller herself. Fox Keller was trained as a physicist but shifted to biology.

She has a long history of discussing how different these scientific practices are. They include differences in the use of modeling, degree of mathematization, and other science practices. My twist for these differences focuses upon the instrumental style of the technics of the two sciences. As the Watson/Franklin narrative shows, the instrument used to discover the *double helix* shape of DNA was X-ray crystallography, the machinery of which is at best "middle-sized," as an X-ray machine fits easily into a single lab. This X-ray imaging technology is a later refinement of often-crude first generation X-ray imaging. If one goes all the way back to early modern science, much argumentation was associated with Galileo's use of optics (both telescopes and microscopes) and the problems of double and triple images, etc. Gradual refinement of clarity is a typical trajectory of improvement in instrument development. In the case of X-ray crystallography, its image, photographed and used in a publication by Franklin (called photo #51, 1953), was the clearest to date, and Watson in one of many scientific "aha" experiences recognized the double-sided helix shape that finally destroyed the previous triple-sided notion of the shape of DNA. My point here is that biochemistry was embodied in a relatively middle-sized instrument, not yet nanoscaled as are many of today's instruments, but easily manipulated by bodily engagement.

This example contrasts strongly with the dominant instrumental shape of what I shall call Cold War physics. Many science historians have noted that in this style of physics, the largest, most complex instruments are needed to image the smallest subatomic particles. The U.S. supercollider, whose history I will cite, was the largest planned supercollider of its time. Its design called for an 87.1-kilometer/54.1-mile tunnel, which contrasts with today's CERN Large Hadron Collider (one of several colliders in the CERN complex). Supported by twenty-two member countries and far too expensive for any one country, including the

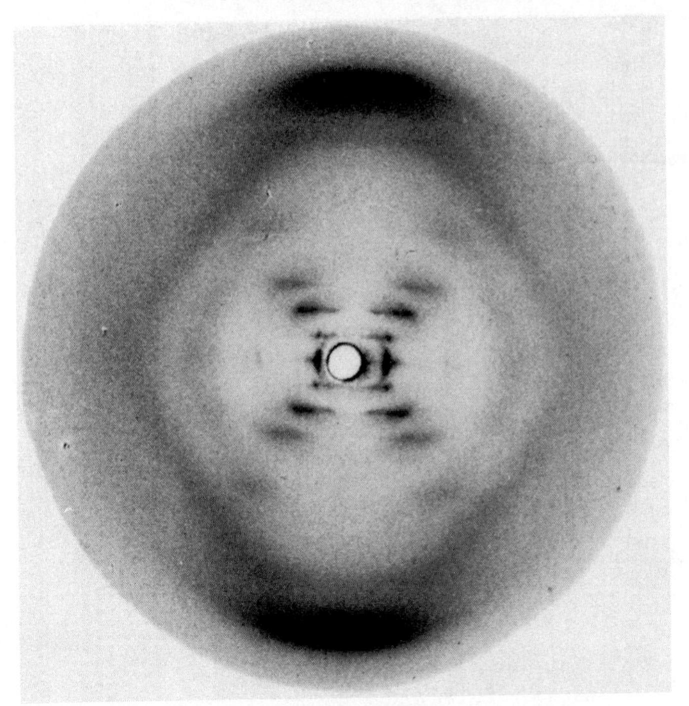

Figure 1. "Photo 51": X-ray diffraction image of crystalline DNA taken by Dr. Rosalind Franklin and Raymond Gosling, 1952. Image copyright King's College London. Used by permission.

United States, today's foremost collider has only a 22-kilometer, very deep tunnel. I mention both as examples of macrotechnics, the style of Big Science physics instruments used to examine the nanoscaled atom particles. Individual parts of this collider, like its magnets, weigh over two tons. Large machines are needed to handle construction and adjustment. The most publicized discovery of the Hadron Collider was the Higg's boson (the so-called "god particle") in 2012. This is the macrotechnics equivalent to Watson and Franklin's discovery of DNA with X-ray crystallography. In my narrative, however, I will focus more upon the U.S. supercollider, whose history is more telling in its 1994 cancellation after the Cold War. This was the largest planned instrumental technics since the emergence of Big Science and the World War II Manhattan Project. The earliest CERN colliders go back to the 1950s, which, I point out, made them roughly contemporaneous with the DNA X-ray crystallography technics. Thus, extrapolating from Fox Keller, by the mid-twentieth century we have physics using macrotechnical instruments to discover subatomic particles and biology using far more microtechnical instruments to probe the secrets of DNA and other microbiological phenomena. Both tended to produce *visual* images in keeping with most scientists' preference.

The U.S. president Dwight Eisenhower has been well noted for his warning after World War II to beware the "military-industrial complex." This warning came from his last interview as he left the office in 1961.[1]

Cancellation of the U.S. Supercollider as Technoscience Crisis

If the invention and development of the Atomic Bomb by the end of World War II was the first Big Science military-industrial project with its constellation of military, engineering, and physics might combined, then the cancelled U.S. supercollider at the end of the Cold War came as a sociopsychological shock to the many participating scientists of the time. Sociopolitically, the "military-engineer-physicist" grouping had long been supported by an eager Congress, which had funded many Big Science

1. Dwight Eisenhower, NPR last interview, April 1961.

projects of the Cold War. Eisenhower's was the era of MAD (mutually assured destruction). But after the collapse of the USSR and the demolition of the Berlin Wall, it appeared that the multibillion-dollar projects fell into disfavor; so after some two years of congressional debate, with worries about vast cost overruns, the U.S. supercollider was cancelled. The reaction from within technoscience was shock. Particularly in physics, which had always gotten the largest share of governmental support, worries began to expand that this was minimally a disciplinary crisis, maximally a signal that the position of physics as the most favored science was doomed.

I experienced this myself, working out of this reaction from within a dominantly math and physics research university, Stony Brook. As a long-term member of the interdisciplinary University Research Committee, I saw the often frantic push for physicists and many related engineering research faculty to shift to biological and medical applications that rapidly grew multimillion-dollar grants and proposals. For example, new prostheses, new imaging, and often engineering-physics-based programs associated with Stony Brook's large medical school emerged as some of the largest projects of university research. Soon multimillion-dollar grants began to be won. On an even larger scale, also going back to my graduate school days at Boston University in the mid-sixties, Charles Delisi trained in engineering and physics, later to become Dean of Engineering at Boston University. Delisi was a deep believer in Big Science and a major player in turning biological research Big. During the build up toward the Human Genome Project, he became one of the primary proponents who argued that biology would never become a major science until it became Big Science, and with this, he argued intensely for the Human Genome Project, which would propel biology into its first multibillion-dollar research project. There was a period of intense internal argumentation within biological science, but eventually as history has it, the Human Genome Project successfully made biology into a Big Science.[2] Of course, biology had already turned to a plethora of research projects that entailed micro- and nanoscaled technics. DNA, biotechnology, and genetic engineering are all outcomes in this context.

What constitutes dominant technics? One measure would be relative funding totals for research by discipline. And this measurement shows that, first, most funding levels have been flat for some time, but there has been a shift from a group of natural sciences, including computer and math, physics, chemistry, and engineering. Natural science totals in the United States run about 30 percent, with nearly 60 percent now going to the life sciences (the National Institutes of Health has always been bigger than the National Science Foundation; the National Endowment for Humanities includes medical research), with the social sciences at a miserly 3.6 percent—there are no statistics for philosophy! Although my argument here is that the biological sciences have clearly gained prominence since the mid-twentieth century, it is also clear that the macro technosciences have not disappeared. In recent articles, I have often taken accounts of what I call *shelf lives,* or time frames in which particular technologies, scientific objects, and even philosophies have finitely useful lives; but just as each can have a beginning, endings also occur. For example, some of our oldest technologies are tools, my favorite of which is an Acheulean hand ax. This bifaced stone tool, which most likely served as a sort of Swiss Army knife for *homo erectus,* first appeared roughly 1.8 MYA (million years ago) and, in what archeologists recognize as the oldest and longest lasting shelf life, was produced

2. The Human Genome Project was completed April 2003 at a cost of $3 billion.

and used in virtually the same manner until roughly 400,000 BP (before present), thus displaying a 1.4 million-year shelf life. It was multipurpose, used for cutting, scraping hides, digging, and possibly even as a thrown weapon. We are all aware that since that time technological artifacts have come to have much shorter or "sped-up" shelf lives. I have often published comparisons, including the roughly similar and parallel shelf lives of typewriters and steam trains at roughly 1.25 centuries each or today's cellphones with shelf lives of only a few years. But equally as important, contemporary shelf life technologies—such as cell phones—are much more widely used than any previous technologies. Many social scientists claim that 95 percent of the entire human population has access to cell phones, probably the most ubiquitous technology ever (many cell phones are owned by a single person who then rents or lends out usage times to entire villages). Note, too, that many contemporary communication technologies are also "Swiss Army" multipurpose: phones, texting, camera, internet, and on and on as with the more ancient Acheulean hand ax. This latter feature is miniaturized in cell phones—and in a whole series of miniaturized technologies such as drones and robots—such that surveillance, observational, spy, and distance sensing reach into mini-satellite, deep sea sensing, and, in acoustic forms, subsurface imaging. Such undersea, satellite, and remote sensing, still dominantly visual, are often preferred at micro-levels for economic and sometimes more secretive uses. Moreover, given funding, it is no surprise that military—but also medical—miniaturization is common. Bee-sized drones can occupy hidden spots for surveillance; cell phone–adapted technologies fill mini-satellites.

In my earlier chapters that take note of my own experience of such medical technics, the technics were also miniaturized. Laparoscopic surgical instruments and technologies such as X-ray, MRI, PET, and GI imaging all rely on a range of magnetic, radioactive, camera, and other microtechnics for both diagnosis and treatment. So far, I have missed using swallowable capsule-sized camera technics, some of the most nano-diagnostic technics, and very avant-garde prescription nano-placement technics. "Aging Cyborg: I Don't Want to Be a Cyborg, I and II" does survey some newer acoustic technics, more elaborately referenced in my 2015 *Acoustic Technics.* Similarly, in my aging section I have not mentioned the proliferation of self-monitoring wearable data technologies that for some people are addictive. These include data generation of everything from steps taken to heart rates and a range of bodily aimed biological phenomena. Then for special diagnoses, there are blood glucose machines and a plethora of other self-collecting data streams. In Japan the proliferation of gerontological robotics include vital sign smart toilets, bathing machine robots, and even talking therapy robots. Many are also microtechnologies worn on the body and usually visually imaged. In my case, similar devices are common in recovery settings, and I have, in experiencing data for blood pressure, pulse, and multiple other rate displays, quickly learned to self-change rates by my own bodily feedback learned reactions. Wearable technologies, of course, reduce time factors that, if in the form of printed lab test results, are much delayed. Some of my readers are dismayed that I do not use social media—Facebook, Twitter, etc.—but the reason is mostly related to sheer screen time demands. Since internet research, email, and news take up inordinate time already, to add social media, particularly with its built-in nudging to self-display, voyeurism, and other exhibitionism, let alone the occasional data breach, does not fit well with my notion of myself.

Postphenomenological Prognosis

Multiuse and Multistability

I have discussed two prominent Swiss Army knife technologies here. The Acheulean hand ax was in use for more than a million years as digger, cutter, scraper, and thrown weapon. And today, the cell phone, the most ubiquitous technology ever, is even more multipurpose as acoustic phone, text sender, camera, photo album, web browser, and so on, with a technological modeling such that its variants range from space probes to deep sea explorers. Postphenomenology understands all technics to be multistable and multiuse, which makes predictability difficult, if not impossible. This variability happens regardless of size or complexity.

Prostheses and Perceptual Multidimensionality

While medical technics abound in prostheses—all of which relate to the ranges of perceptual experience, from the tactile-kinesthetics of Merleau-Ponty's examples to Nintendo surgery to my hearing aids and cataract lenses—all condense upon the whole-body motility that determines any phenomenologically related sense of embodiment. I have noted here how both art and science, through embodying technics, are fully technoart and technoscience in action.

Technologies

From early modern times until now, technologies have stretched our mediated experience to dimensions unknown and never experienced in antiquity. This may be most easily discerned at both macro- and microlevels. Much of what we can now mediatedly experience was simply unknown prior to the twentieth and twenty-first centuries. In astronomy, as I have frequently pointed out, until the accidental discovery of radar, all astronomy was "white light" limited. Today we can image and thus mediatedly experience the full range of the electromagnetic spectrum, from radio waves down to gamma rays. The same happens in medical technics. We now have multiple imaging perspectives on interiors: X-rays, PET scans, MRI and fMRI. Sonograms, both static and dynamic, and frequently composite modeled through computer tomography, can vividly show everything from fetuses to brain tumors or, as in the consultation before my heart surgery, a normal multiscreen display in full, glowing color.

Animal Studies Are Indirectly Related to Medical Technics

Here the limits of human perception have again been breached. Infra- and ultra-dimensions in virtually all sensory realms have become technically and mediatedly available: thermal, magnetic, infra- and ultra- sound and vision are included in documentaries for all to appreciate. I vividly remember my first experiences of whale songs and throat singing. And the first sounds of ultrasonic male mice songs and the responding crooning of female mice.

Similarly, today's imaging of the vast internal colonies of microorganisms amaze.

All this says much about how different our world is from the worlds of our early human ancestors. And yet, even as we uncover—through the jungles and overlays of the earth—we can sense, too, that we are simply beginning to expand this world that has yet to show its still greater complexity.

We Make Technology, Technology Makes Us

Interview with Daisy Alioto[1]

Philosopher Don Ihde discusses a human future shaped by the technology we've created.

MEDICAL IMAGING is responsible for some of the most powerful moments of our lives, from the first glimpse of a growing fetus to the discovery of a tumor. Before the invention of the X-Ray in 1895, none of it was possible, but imaging is now a fundamental part of diagnosis and even how we understand our selfhood. If we considered these images to be art, then philosopher Don Ihde would be one of their most prolific collectors.

Ihde is a distinguished professor of philosophy at New York's Stony Brook University. He is the author of over twenty books, including the 1976 *Listening and Voice: A Phenomenology of Sound*—the first and only phenomenological investigation of the experience of sound—and a follow-up, 2015's *Acoustic Technics*.

As a phenomenologist, Ihde studies consciousness. Specifically, he's a postphenomenologist, meaning he believes we can only interpret our experiences through the confines of our own brains—and the audio and visual instruments that mediate them. Simply put, Ihde studies the philosophy of technology. His next book will focus on how technology and instruments are now shaping

1. Originally published on GuernicaMag.com 12/1/2017.

the humans of the future just as they shape our understanding of the past. Over a root beer float on the Upper East Side, we talked about echolocation, Vermeer, and Lucy's skeleton.

—DAISY ALIOTO for *Guernica*

GUERNICA: I'm interested in how you started exploring medical imagery and what types of media you're seeing it expand into.

DON IHDE: A lot of the medical imagery has to do with biography because I had open-heart surgery, I had knee replacements, I had a hiatal hernia, etc. Every time one goes for surgery you get a whole spectrum of imaging. Of course, I've been doing research in imaging technology across the board for close to twenty years. When you think about it, medical imaging is actually quite new. The first major medical image was the X-Ray in 1895. This is the first time you get imaging of anything that's in the bodily interior.

GUERNICA: Versus drawings.

DON IHDE: Yeah, Australian Aborigines have a style of X-Ray painting. In the European tradition, you go back to the Renaissance when you start doing autopsies. But to peer inside a living body—X-Rays are the earliest. Now of course, it's absolutely amazing. I have probably a dozen CDs of my own imaging, and a couple of times I've used them in lectures. At Oxford, I did a presentation called "My Case." What specialists try to do is get at least three imaging processes that are totally different from each other. Then you can run these through a computer program and make a composite image. In one scenario you suspect a brain tumor, so you image the brain tumor with PET scans, MRIs, and CT scans and create a 3-D model. The doctor opens up the skull to excise the cancer, but they can't see anything. Do you cut out what's supposed to be in that spot or not?

The current story is yes, you believe the images over what you see with your eyes.

GUERNICA: Is that an actual case, or is it a medical school scenario?

DON IHDE: It's a sort of urban legend, but the notion is true.

GUERNICA: The trust in medical imaging is that great?

DON IHDE: Yes. There are several books on the history of diagnosis, and that's very interesting because, gender-wise, it had long been prohibited in Asia and Europe for male doctors to touch females. So there are periods in history when they created dolls, and the woman would point to the part of the doll where the pain was. Later, post-Renaissance, that taboo is gone. A lot of the probes were combined tactile and acoustic. For example, a doctor might thump an abdomen to see if there is a tumor. You have a long history of changes in diagnosis, from no touch, to touch, to acoustic and visual. But visual is a problem because if you're a living being, you can't see beyond the surface of the skin. Now you don't have that problem. Laparoscopic surgery, inserting a camera into the body, is sometimes called Nintendo surgery, and the best training for laparoscopic surgeries have actually been video games. A developing field is sonifying cancer. You can acoustically discriminate between healthy and cancerous cells by hearing them. Make the vibrations into sound, and anybody can learn to tell the difference.

GUERNICA: When you characterize the evolution in diagnosis from visual to tactile to acoustic and tactile, do you see imaging as a return to visual, or do you see it as hypervisual?

DON IHDE: In the eighteenth century, it was very common to think that each of the five senses was discrete and yielded very different data. That period is totally dead. A Jesuit priest named

Lazzaro Spallanzani was the first to discover bat echolocation. He saw that bats could fly in the dark and catch a moth. He put wax over the bat's eyes but the bats could still catch the moth. If you put wax in the bat's ears, the bats couldn't catch anything. So I argue that science would be much richer if it were multisensory. And the problem with instrumentation is that instruments, unlike our senses, can be monosensory. Since the nineteenth century and the discovery of the electromagnetic spectrum—which is really the discovery that all energy coming from something has a wave form—in theory we could image anything along that spectrum. In fact we don't, because only certain parts of the spectrum have been instrumentalized. But the new thing is computerization. You can take all the data, the measurement of the frequencies, and transform it into an image.

GUERNICA: Coming at it from the perspective of a collector or someone who appreciates art, do you foresee a scenario where people will begin to collect these images or audio samples, and they'll be decontextualized from the medical field?

DON IHDE: When I give lectures on sonifying science, many times there will be artists in the audience who give me samples of stuff they've done. There are annual contests put on by some of the top science magazines for best science images. They range from regular photography to the phenomenon of the microphotography of cells. Twenty-first-century imaging is largely using microscopic processes. So, for example, you can trap a single atom or a single proton. Two years ago, one of my colleagues was the fourth member of an atom trapping consortium. Unfortunately, Nobel Prizes can only be split by three, so he was the guy who was left out.

GUERNICA: You've kept all of the images from your own cases.

Do you keep them because they're useful to your work, or do you have an instinct to save them for other reasons?

DON IHDE: I originally kept them because I wanted to use them as examples in what I write, and I do, a lot, but it's very interesting because we have no direct experience of our brain. I can't experience my brain because I'm inside of it. If you're imaging your brain, you can also find scary things. As one ages, your brain shrinks. And how much it shrinks, and where it shrinks, relates to conditions like Alzheimer's and dementia.

GUERNICA: Do you think that having recordings and imagery that help us experience our own brains has a philosophical impact?

DON IHDE: It clearly has. My thesis with *Philosophy of Technology: An Introduction* is that we make technologies, and, in turn, technologies make us. There is a notion in design of designer intent. Why was the lead pencil invented? Well, the inventor of the lead pencil wanted it to be a marking machine. The dominant use of a pencil is to write, but I remember going to a one-room public school, and one of my bullying friends stabbed me with a lead pencil. I remember reading about a court case where a man tried to stab a judge with a pencil. There are Google pages full of similar instances around the world. It's obvious that the pencil lends itself to precisely that kind of use. It's not as lacking in dominance as you might think. I have an article on the fallacy of the designer intent because a lot of designers think they can design uses into technology. You can't do that. I use the pen, I make the mark, but the pen is also using me. The pen could be said to be allowing these kinds of marks. I can't do just anything with the pen. That's particularly true of imaging.

GUERNICA: Instruments play a big role in this.

DON IHDE: The interrelationship is true from the simplest to the most complex. It doesn't make any difference whether it's art or science.

GUERNICA: There were artists who, even before computers, were creating their own coding systems to create analog works based on some ratio, or what we would call a program if it was in a machine. Computer art is really not that new—whether it was the Fibonacci sequence, or whatever other system.

DON IHDE: In the early 2000s David Hockney produced a book with Charles Falco called *Secret Knowledge*. It's about his discovery that many Renaissance painters used a camera obscura to construct paintings. Many in the art history community were outraged by this because they thought it cheapened the art. I happen to be a big Vermeer fan. There can be little doubt that he used a camera obscura. He's only got something like thirty-nine paintings and many of them show exactly the same room.

GUERNICA: So that's why it was so funny to you, the uproar about the camera obscura.

DON IHDE: I never went into aesthetics. Aesthetics is what philosophers have to say about art, and a lot of them take an analytics position and raise the question, "What is an art object?" As soon as you fall into that trap, an artist is going to come along and say, "That isn't art it's something else." That's a hopeless gig.

GUERNICA: But what about defining beauty? Is it harder to rewrite a definition of beauty?

DON IHDE: I don't like definitions at all.

GUERNICA: So you couldn't be a lawyer.

DON IHDE: Definitions get you into that time trap, and I'm very much more process-focused. Take Lucy for example. Lucy is famous largely because she has almost a total skeleton. The more sophisticated we get with instruments, the more we can find out. Through CT scans of her skeleton, they now think she died falling out of a tree because of the way her bones are broken. If nineteenth- and twentieth-century technologies can retroactively transform our bodiment, what then do the technologies we now use do?

Additional Resources

Ihde, Don. *Embodied Technics*. Copenhagen: Automatic Press, 2010.

Ihde, Don. *Expanding Hermeneutics: Visualism in Science*. Evanston, Ill.: Northwestern University Press, 1998.

Ihde, Don. *Ironic Technics*. Copenhagen: Automatic Press, 2008. See especially chapters 3 and 4.

Friis, Jan Kyrre Berg O. and Robert P. Crease, eds. *Technoscience and Postphenomenology: The Manhattan Papers*. Lanham, Md.: Lexington Books, 2015. This book collects papers given at a retirement conference in March 2012 at Stony Brook Manhattan.

Rosenberger, Robert and Peter-Paul Verbeek, eds. *Postphenomenological Investigations: Essays on Human–Technology Relations*. Lanham, Md.: Lexington, 2015. Similar case studies by other postphenomenologists with an overview of postphenomenology by the editors.

Selinger, Evan, ed. *Postphenomenology: Critical Companion to Don Ihde*. Albany, N.Y.: SUNY Press, 2006. This was the first *Festschrift*-style book. Following were two books, mostly with essays related to postphenomenological case studies and in the Lexington Books "Postphenomenology and Philosophy of Technology Series."

Postphenomenology now has a series, "Postphenomenology and the Philosophy of Technology," with Lexington Books, an imprint of Roman & Littlefield. To date several books relate to various technologies discussed in Medical Technics, *including the following:*

Ihde, Don. *Acoustic Technics*. Lanham, Md.: Lexington Books, 2015.

Verbeek, Peter Paul. *What Things Do*. University Park: Pennsylvania State University Press, 2005.

Wellner, Galit. *A Postphenomenology of Cell Phones*. Lanham, Md.: Lexington Books, 2015.

Articles, often presented at STS conferences, have also been published, including the following:

Besmer, Kirk. "Embodying a Translation Technology." *Techne* 16, no. 3 (2012): 296–316. On cochlear implants.

De Preester, Helena. "Technology and the Body: The (Im)Possibilities of Re-embodiment." *Foundations of Science* 16, no. 2 (2011): 119–37. On prostheses.

Olesen, Finn. "Technological Mediation and Embodied Health-Care Practices." In *Postphenomenology: Critical Companion to Ihde*, edited by Evan Selinger, 231–46. Albany, N.Y.: SUNY Press, 2006.

Rosenberger, Robert. "A Case Study in the Applied Philosophy of Imaging: The Synaptic Vesicle Debate." *Science, Technology, and Human Values* 36, no. 6 (2011): 6–32. On slam freezing.

Rosenberger, Robert. "Mediating Mars: Perceptual Experience and Scientific Imaging Technologies." *Foundations of Science*, no. 18 (2013): 25–31. On distance sensing.

Acknowledgments

First, to my best critic-vetters who know what they did, Cathrine Hasse, Robert Rosenberger, and Robert Scharff, to whom I also dedicate this volume; then to Linda, my wife, who helps with my computer problems and gives advice and help; then to the University of Minnesota Press team: Danielle Kasprzak, Anne Carter, Doug Armato, and the others unnamed; to the whole Stony Brook group, colleagues, especially Robert Crease, Marshall Spector, Lorenzo Simpson, and to my six bright PhD advisees doing phil-tech dissertations from 2006 to 2013, plus many more from other countries; the stimulation of the techno-science research seminar with its quadruple continent visitors: Asia, the Americas, and Europe.

(Continued from page iii)

Ian Bogost
The Geek's Chihuahua: Living with Apple

Shannon Mattern
Deep Mapping the Media City

Steven Shaviro
No Speed Limit: Three Essays on Accelerationism

Jussi Parikka
The Anthrobscene

Reinhold Martin
Mediators: Aesthetics, Politics, and the City

John Hartigan Jr.
Aesop's Anthropology: A Multispecies Approach

Don Ihde is distinguished professor of philosophy emeritus at Stony Brook University, New York. His books include *Bodies in Technology* (Minnesota, 2001) and *Husserl's Missing Technologies.*